Evaluation of Occupational and Environmental Exposures to Radon and Radon Daughters in the United States

Recommendations of the
NATIONAL COUNCIL ON RADIATION PROTECTION AND MEASUREMENTS

Issued May 31, 1984

National Council on Radiation Protection and Measurements
7910 WOODMONT AVENUE / BETHESDA, MD 20814

Library of Congress Cataloging in Publication Data

National Council on Radiation Protection and Measurements.
Evaluation of occupational and environmental exposures to radon and radon daughters in the United States.

(NCRP report, ISSN 0083-209X ; no. 78)
Bibliography: p.
Includes index.
1. Lungs—Cancer—United States—Etiology. 2. Radon—Toxicology. 3. Uranium mines and mining—Hygienic aspects—United States. 4. Environmentally induced diseases—United States. 5. Occupational diseases—United States. 6. Radiation dosimetry. I. Title. II. Series.
RC280.L8N37 1984 363.1'79 84-4756
ISBN 0-913392-68-5

Preface

It is well known that occupational exposure of underground miners to high levels of radon daughters has resulted in radiation-induced lung cancer. Alpha-emitting radon daughter exposure also delivers the highest dose from background radioactivity to the global population. The concern over the "natural" radon daughter exposure has been increased by the results of various programs which have demonstrated "enhanced" radon daughter levels in dwellings around the world. Directly estimating the effect of environmental levels of radon daughters has not been possible due to the normal incidence of lung cancer complicated by other cancer-causing agents such as smoking and environmental pollution. Using the mining experience to estimate the effect of the "enhanced" levels of radon daughters on the general population has been difficult because of the problems involved in extrapolating from the relatively homogenous occupational group to the highly variable general population.

This report considers the sources of radon and the dosimetry of radon daughters in the mine and the general environment. The variability of the radon daughter dose conversion factor is examined in terms of the miner, as well as the adult male and female, child, and infant in the general population. The adequacy of the working level (WL) as an exposure unit is evaluated. Lung cancer in man and experimental animals is examined.

The report describes a model for estimating the probability of lung cancer from exposure to radon daughters at the environmental levels of concern to the general population as well as at the levels involved in occupational exposure of the miner. It is based upon the most recent estimates of lung cancer in underground uranium miners, and accounts for the apparent increase in lifetime risk with increasing age at first exposure noted in epidemiological studies of underground uranium miners.

It is not definitely known if the extrapolation from the occupational experience to the general environmental situation is valid. However, it is consistent with the present radiobiological concept that lung cancer induction is a stochastic process without threshold.

The potential risk from the present radon daughter exposure standard for miners is calculated, as well as estimates of risk to the general population from an assumed average environmental level of radon daughters.

The Council has noted the adoption by the 15th General Conference of Weights and Measures of special names for some units of the Système International d'Unités (SI) used in the field of ionizing radiation. The gray (symbol Gy) has been adopted as the special name for the SI unit of *absorbed dose*, *absorbed dose index*, *kerma*, and *specific energy imparted*. The becquerel (symbol Bq) had been adopted as the special name for the SI unit of *activity* (of a radionuclide). One gray equals one joule per kilogram; and one becquerel is equal to one second to the power of minus one. Since the transition from the special units currently employed—rad and curie—to the new special names is expected to take some time, the Council has determined to continue, for the time being, the use of rad and curie. To convert from one set of units to the other, the following relationships pertain:

$$1 \text{ rad} = 0.01 \text{ J kg}^{-1} = 0.01 \text{ Gy}$$
$$1 \text{ curie} = 3.7 \times 10^{10} \text{ s}^{-1} = 3.7 \times 10^{10} \text{ Bq (exactly)}.$$

This report was prepared by Task Group 4 (Radon and Daughters) of Scientific Committee 57 on Internal Emitter Standards. Serving on the Task Group were:

Naomi H. Harley, *Chairman*
Institute of Environmental Medicine
New York University Medical Center
New York, New York

Members

Fred T. Cross
Battelle Pacific Northwest Laboratory
Richland, Washington

Bruce D. Stuart
Stauffer Chemical Co.
Farmington, Connecticut

Advisors

Victor E. Archer
University of Utah
Salt Lake City, Utah

Donald A. Morken
University of Rochester
Rochester, New York

Consultant

John H. Harley
Environmental Consultant
Hoboken, New Jersey

Serving on the Scientific Committee were:

J. Newell Stannard, *Chairman*
University of California
San Diego, California

John A. Auxier
Applied Science Laboratory
Oak Ridge, Tennessee

William J. Bair
Battelle Pacific Northwest Laboratory
Richland, Washington

Patricia W. Durbin
University of California
Berkeley, California

Keith J. Eckerman
Oak Ridge National Laboratory
Oak Ridge, Tennessee

Roger O. McClennan
Lovelace Inhalation Toxicology Research Institute
Albuquerque, New Mexico

Paul E. Morrow
University of Rochester
Rochester, New York

Robert A. Schlenker
Argonne National Laboratory
Argonne, Illinois

Roy C. Thompson
Battelle Pacific Northwest Laboratory
Richland, Washington

NCRP Secretariat—**E. Ivan White**
James A. Spahn

The Council wishes to express its appreciation to the members, advisors and consultant for the time and effort devoted to the preparation of this report.

Warren K. Sinclair
President, NCRP

Bethesda, Maryland
April 15, 1984

Contents

1. Summary

The radon daughter dose which is most significant for any population exposure, occupational or environmental, is the alpha dose received by the tracheobronchial (TB) region through inhalation of the airborne, short-lived daughters ^{218}Po (RaA), ^{214}Pb (RaB), ^{214}Bi (RaC) and ^{214}Po (RaC′). Two radionuclides, ^{214}Po (RaC′) and ^{218}Po (RaA), deliver the alpha dose that is discussed in detail in this report. The complexity in the dose estimates for the lung required to account for daughter deposition, radioactive build-up and decay, removal by mucociliary clearance, and physical dose calculation for specific cells in bronchial mucosa, has caused many to consider a dosimetric approach unduly difficult. However, lung cancer induced by radon daughters in underground miners is found primarily in the upper TB region. Most of the modeling parameters that are necessary for assessment of dose in this region are reported in the literature and for this reason, it is now possible to estimate the dose to this region rather than to the whole lung, so that a proper dose-response relationship may be formulated.

Historically, exposure is defined in terms of the air concentration of radon daughters in units of the working level (WL). This is defined as any combination of short-lived daughters in one liter of air that will result in the emission of 1.3×10^5 MeV potential alpha energy. This avoids the problem of degree of radioactive equilibrium among the daughters. Cumulative exposure is documented in working level months (WLM) which equals exposure in working level times exposure duration in multiples of the 170-hour occupational month. These special units and how they are utilized are described in detail in Section 7. The absorbed alpha dose to cells in the bronchial epithelium in the upper airways of the tracheobronchial tree is the significant dose for cancer induction. In this report, the absorbed alpha dose to cells per unit cumulative exposure (dose conversion factor) is given in units of rad/WLM. The dose per unit cumulative exposure is derived for both occupational and environmental conditions in Section 4 and 5 and the overall variability of the dose conversion factors is discussed in detail in Section 6.

The underlying consistency in the human epidemiological data, when lung cancer is related to exposure in WLM, is probably due to the relatively narrow range of bronchial dose per WLM. The dose conversion factors have been calculated by many investigators for persons exposed to radon daughters occupationally (underground mining) and environmentally. This report adopts the alpha dose to basal cells in bronchial epithelium located in generation 4 in the TB tree (the site of lung cancer in miners), at a depth of 22 μm below the epithelial surface where shallowest basal cells are located. The values of the dose conversion factor adopted in this report are:

	rad/WLM
Underground Miners	0.5
Environmental Exposures	
Men	0.7
Women	0.6
Children	1.2
Infants	0.6

There can be substantial changes in the magnitude of this dose conversion factor (a factor of two) due to differing characteristics of the inhaled atmosphere: these are not accounted for when using the working level month as a unit of exposure. While past exposures cannot generally be reconstructed and a single conversion factor must suffice, it is recommended that future measurements should provide the aerosol data to allow calculation of the specific bronchial dose as well as documenting exposure in WLM.

The source of radon (and thus radon daughters) in the atmosphere is radium present in the earth's crust, in water and in building materials. In solids, radon atoms can escape only if the decay occurs within recoil distance of pores or other open space, while in water, they are held only by solubility. Unlike many manmade materials, where radon release is very low, most soils have the ability to release more than 10 percent of the radon formed. The fraction of radon released (emanating power) from a solid material will depend on its porosity, but even more on the history of the material which controls whether radium is on or near the surface of particles. It is currently believed that the source of most of the radon indoors is soil beneath the dwelling. The structure itself acts merely to contain radon released from soil. Ventilation can reduce any indoor radon concentration effectively to outdoor levels with about four air changes per hour. As this high ventilation rate is seldom attained, indoor levels are generally considerably higher than outdoor levels and are a significant contributor to any population dose.

The sources of radon on a global scale can be estimated from existing data (other than the contribution from all building materials which is not known at present) and these annual releases for the three major sources are:

	Ci per Year
Soil	2×10^9
Ground water (potential)	5×10^8
Oceans	3×10^7

This annual release of about 2.5×10^9 curies of radon leads directly to a global atmospheric inventory of 3.6×10^7 curies, as there are no known sinks. This can be converted to an average surface air concentration of 70 pCi/m^3, or about 200 pCi/m^3 over the continents.

The animal studies to determine the effects of inhaled radon daughters have been supportive of the human epidemiological results. Five similarities are most apparent.

1) Tumor production per WLM at very high exposures (>1000 WLM) and exposure rates to radon daughters is much lower than at moderate exposures. This has now been seen in hamsters, rats, and dogs as well as man. The lowest attributable lung cancer rates per unit exposure were observed in the U.S. uranium miners and in the high exposure cohort in Canadian fluorspar miners.
2) There is some preliminary evidence that long exposures at lower dose rates are more productive of lung tumors. This is seen in rats, where the percentage of tumors per WLM was highest at the lowest doses and dose rates. In humans, the highest risk coefficient calculated (50×10^{-6} lung cancers per year per WLM) is for persons exposed later in life. An average risk coefficient obtained for all exposure categories and all age groups of 10×10^{-6} lung cancers per year per person per WLM is adopted in this report.
3) A lower lifetime incidence of lung cancer is observed in dogs exposed to cigarette smoke along with radon daughters rather than to radon daughters alone. The effect was also observed in a small group of Swedish zinc-lead miners, and is tentativelyascribed to the protective effect of increased mucus production from smoking. Swedish miners at Malmberget did not show this effect, however, and the effect of smoking upon radon daughter induced lung cancer is uncertain.
4) Emphysema can be attributed to radon daughter exposure in both animals (hamsters, rats and dogs) and underground miners.

The presence of ore dust or diesel fumes does not appear to be a prerequisite for lung cancer.

5) The predictions of the various dosimetric models appear to be borne out in the various species. The tumors induced in studies- with hamsters and rats, which have similar lung morphometry, occur in the distal portion of the conducting airways or in the pulmonary region. These regions receive the highest dose based upon calculations (~ 0.5 rad/WLM). Human tumors appear almost exclusively in the upper airways of the bronchial tree. Absorbed dose calculations show that basal cells in the upper airways and particularly the fourth generation or segmental bronchi is the location in the human respiratory tract that receives the highest dose from radon daughters (0.5 rad/WLM for miners).

A model for predicting lung cancer deaths from exposure to radon daughters at environmental levels has been developed for this report. It is based upon most recent estimates of lung cancer in underground uranium miners and accounts for the apparent increase in lifetime risk with increasing age at first exposure noted in epidemiological studies of underground uranium miners. Although the model appears to represent the uranium miner lung cancer response well, it is not known whether extrapolation to environmental levels is valid. Extrapolation to environmental levels is adopted in this report, however, since it is consistent with the present radiobiological concept that lung cancer induction is a stochastic process without threshold. Studies in 18 cities in Canada (LeTourneau *et al.*, 1983) could not detect a relationship between radon daughter concentrations and lung cancer mortality rates. An average risk coefficient, obtained for all exposure categories and all age groups, of 10×10^{-6} lung cancers per year per person per WLM is adopted in this report. This corresponds to a lifetime risk of about 1.5×10^{-4} per WLM but is naturally dependent upon exposure duration and age at first exposure. For comparison ICRP (ICRP, 1981) has adopted a range for lifetime risk of $1.5\text{–}4.5 \times 10^{-4}$ per WLM.

Using this approach and the adopted risk coefficient of 10×10^{-6} lung cancers per year per WLM, an estimate of the risk to individuals involved in exposing a population to radon daughters through a given practice may be obtained. Environmental levels are frequently given in terms of the radon concentration, so the lifetime risk for continuous environmental exposure to 1 pCi^{222} Rn per m^{3} with 70% daughter equilibrium for different time intervals is summarized in the following table. This summary table may also be used to estimate the additional or incremental risk if additional ^{222}Rn is introduced into an environment due to energy saving or other practices. For purposes of comparison, the lifetime risk of lung cancer through exposure to the average outdoor ^{222}Rn concentration over continents of 200 pCi/m^3 is 0.0004 or 0.04 percent, based on the risk estimates in the summary table.

Exposure Duration	Lung Cancer Risk per pCi $^{222}Rn/m^3$ (Environmental Conditions)
1 year	5.1×10^{-8}
5 years	2.6×10^{-7}
10 years	5.1×10^{-7}
30 years	1.4×10^{-6}
Lifetime	2.1×10^{-6}

An evaluation of the levels of radon daughters which yield a lifetime risk of lung cancer in occupational and environmental settings conclude this report.

2. Introduction

Alpha-emitting radon daughter exposure to the lung delivers the highest dose from background radioactivity to the global population. It is well known that exposures of underground miners to high levels of radon daughters have produced the largest numbers of radiation-induced lung cancers found in any exposed population. These cancers are found primarily in the tracheobronchial tree rather than the pulmonary parenchyma.

The occupational standard for radon daughter exposure in the United States is currently set at 4 working level months per year (4 WLM per year). Average environmental exposure in the U.S. is thought to be about 0.2 WLM per year. Adequate data do not exist to give a good estimate of the true range. Thus, the entire population of the U.S., if not the world, is exposed to approximately 5 percent of the permissible occupational value, and some individuals are undoubtedly exposed to many times this value depending upon geographic location.

The alpha dose delivered to the lungs of individuals will differ somewhat per unit concentration of radon in the atmosphere because of differences in the characteristics of various atmospheres (aerosol particle size, radon daughter ratios, etc.) and because of variations in radon daughter deposition in the lung with breathing rate, and bronchial morphometery (sex and age differences). The absorbed dose to the tracheobronchial epithelium is, therefore, the underlying common factor which should be used to compare exposures of groups of people.

It is the purpose of this report to examine the sources of radon daughters in the United States, indicating areas of possible unusual exposure, and to examine the data that exist concerning the magnitude of the alpha dose delivered per unit exposure (referred to in this report as the dose conversion factor) for persons exposed occupationally and environmentally. It is also the purpose of this report to examine both human and animal dose response data on lung cancer and to determine whether the experimental data with animals are qualitatively similar to, or supportive of, the human experience; and to estimate the average lung cancer risk to exposed persons. Since exposures are now documented in units of working level months, it is necessary to examine this exposure unit critically to see whether it is sufficiently descriptive for comparison of exposures. The report will indicate how the existing data on experience with high levels of exposure in underground mines may be used to establish guidelines for environmental as well as occupational exposures.

3. Sources of Radon in the Atmosphere

3.1 Introduction

The true source of radon and its daughters in the atmosphere is the radium present in various materials such as soil, water, and building materials, since the amount of airborne radium is negligible. Gaseous radon-222 with its 3.8-day half life can move independently over great distances and maintain a significant concentration in the atmosphere for days even when separated from its parent (see Fig. 3.1). On the other hand, the short-lived radon daughters, which are the radionuclides of concern in human exposure, have no independent existence in the atmosphere. Their half life of about one-half hour does not allow them to maintain a significant concentration for long when separated from the parent. Conversely, the daughters approach equilibrium quite rapidly when no separative processes are active. Thus, we tend to look at the radon concentration and its variability as indicators of possible radiation dose to the lung from the short-lived daughters. The degree to which any material contributes to atmospheric radon concentrations depends on the radium content and the fraction of the radon formed in decay that is released to the atmosphere. This section will attempt to indicate the relative contribution of various sources of radon to both global and local concentrations.

3.2 Formation and Release of Radon

Radon atoms formed in rock or in soil particles can escape only if the decay occurs within recoil distance of some open space. Otherwise the radon is trapped within the solid since the diffusion rate in solids is slow compared with radioactive decay. The relatively high fraction of radon which escapes from soils indicates that much of the radium must be on or near the surface of the soil particles. Tanner (1964) reviewed the overall process of radon formation and transport. He noted that radium precipitated with barium sulfate, nitrate or chloride and distributed uniformly throughout the crystal, released much less than one percent of the radon formed for 200-micrometer grains. On

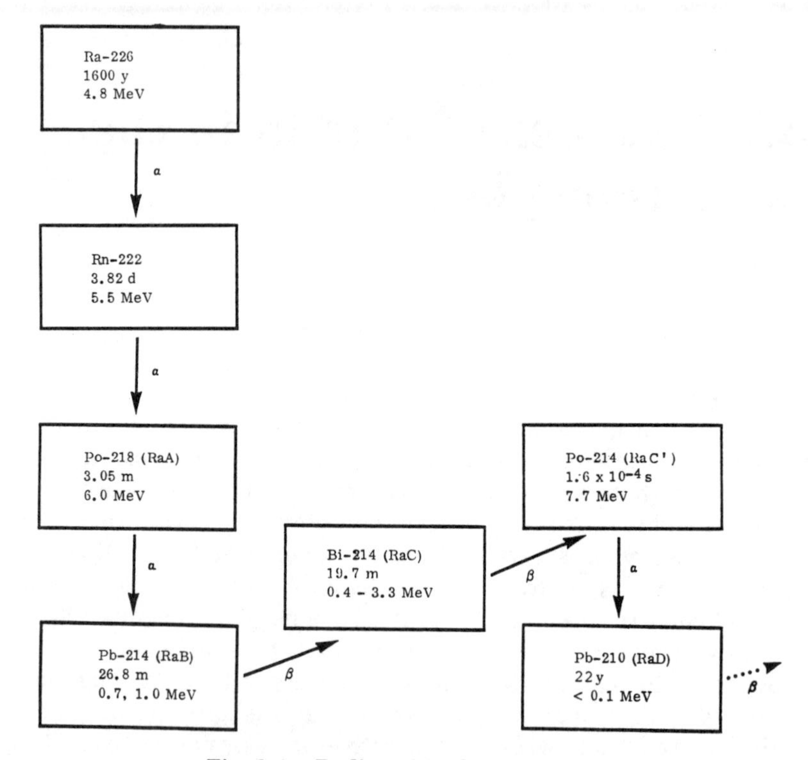

Fig. 3.1. Radium-226 decay scheme.

the other hand, most soils with much larger particle sizes release more than 10 percent of the radon. Thus, the fraction released (emanating power) of a solid material will depend on the porosity, but even more on the history of the material which regulates whether the radium is on or near the surface of particles.

Any radon that escapes from the particles of soil or other solid material enters the gaseous or liquid phase between the particles. Depending on the location and other conditions, some fraction of the radon will be released to the atmosphere.

We might consider the reference radium concentration in normal soil to be 1 pCi/g. This gives us a basis for comparison when considering other materials. A number of studies have been made of the rate of radon emanation from soil and a good rounded value might be taken as 0.5 pCi/m^2/per second. (Birot, 1971; Wilkening *et al.*, 1972). If we were to assume that all of the radon emanation comes from the top meter of soil, this would mean that one-sixth of the radon formed escapes. Since some radon comes from greater depths, a rounded value of 10 percent emanation can be used to describe our reference soil.

Thin layers of soil measured in the laboratory show emanation of up to 50 percent (Baretto *et al.*, 1972), while the relative release from other materials is equal to or less than soils. Massive rock shows little emanation (Baretto *et al.*, 1972) and even small particles of a glassy substance such as fly ash will release less than one percent of the radon formed (Beck *et al.*, 1978). It turns out that most building materials emanate less than soils, and this is further reduced by many of the decorative or protective coatings applied.

Radon that is produced or dissolved in water is held only by solubility. The solubility is sufficient so that emanation at normal temperatures and pressures is small compared with the same area of soil. The radon does follow the laws of gas solubility, however, and can be removed by heating, by reducing the external pressure, or by bubbling another gas through the water.

3.3 Transfer of Radon to the Atmosphere

It is worth remembering that radon produced in a material forms at a steady rate that is independent of any external factors. Thus, in a closed space, the concentration will build up to an equilibrium value dependent on the amount of radium present and the volume. The soil gas volume at depth might be considered such a space, since the amount of radon that emanates from the surface only decreases the concentration near the surface. Kraner *et al.* (1964) showed the following radon profile in weathered tuff, a fairly porous material:

Depth (cm)	*Radon (pCi/m³)*
60	120,000
120	300,000
180	570,000
240	900,000

These values might be compared with the expected atmospheric radon concentration of 100–200 pCi/m^3 estimated from Kraner's emanation rate measurements.

The transport of radon can be by diffusion or by movement of the air or water containing the radon. In the case of radon released into the soil gas, most of the movement occurs by diffusion. The mean migration distance in dry soil is about one meter before the radon decays, while the diffusion rate in free air is considerably faster.

The contribution to air concentration of radon from various depths in soil has been estimated. For uranium tailings, the maximum emanation rate was found with a depth of 3 to 4 meters of tailings and it was estimated that the relaxation length was about 1.5 meters.[1] This would mean that about 50 percent of the radon emanating from the surface would come from the first meter, 75 per cent from the first two meters, etc. The depths involved for a more solid soil containing clay would be about one-half of these values (Kraner *et al.*, 1964).

The effects of various meteorological factors have been well studied. The most noticeable change is that a drop in atmospheric pressure pumps out some of the soil gas which can be at a concentration several orders of magnitude above that in the atmosphere. Continued low pressures cannot maintain this effect since added radon must be brought from greater and greater depths as time goes on. Other meteorological factors do not seem to be very significant and the second major factor in controlling emanation is the state of the soil surface. Snow cover or standing water reduces the emanation rate but there is conflicting evidence regarding moderate amounts of soil moisture. It has been reported (Kraner *et al.*, 1964) that up to 20 percent saturation increases the emanation rate while above that the emanation is retarded. There is also conflicting information regarding the effects of freezing on emanation, but it is most likely that the rate is reduced, e.g., Kraner *et al.* (1964) found that 6 inches of frozen soil reduced the emanation by 40 percent.

The near-surface waters of the oceans contain only about 10^{-4} of the soil radium concentration and the emanation is correspondingly low. Wilkening and Clements (1975) adopted a value of 0.003 pCi/m^2 per second from their measurements. The majority of the radium in the oceans is contained in the sediments, where it cannot contribute to atmospheric radioactivity.

The intimate contact of radon in the soil gas with ground waters leads to some high concentrations in the water. This radon can be released, for example, when hot springs reach the surface or when ground water evaporates. In any case, the ground water can be a transfer mechanism for radon coming from beyond its diffusion range.

The upward movement of radon in the atmosphere is partly by diffusion and partly by atmospheric turbulence. The radon from a particular source will build up a concentration inversely proportional

[1] The relaxation length is the depth of soil which delivers a surface flux that is $1/e$ of that delivered by soil at the surface.

to the volume into which it is emanating. Any confinement tends to produce higher levels, as indicated by the concentrations in soil gas, in mines, in houses, or, for the present case, when the atmosphere is subject to temperature inversions. Conversely, atmospheric turbulence tends to mix the radon into a larger volume and thus reduce the concentration.

Surface winds will carry radon horizontally from the source, but this will have little effect on atmospheric concentrations where a large land mass is contributing radon fairly uniformly. The emanation rate from oceans or other large bodies of water is very low, so that on-shore winds tend to reduce atmospheric radon concentrations in coastal areas. The order of magnitude may be indicated by the generalizations in the report of the United Nations Scientific Committee on the Effects of Atomic Radiation (UNSCEAR, 1972). They listed radon concentrations of 100 pCi/m^3 in continental air, 10 pCi/m^3 in coastal areas and islands, and 1 pCi/m^3 over oceans and arctic areas.

The conditions described lead to marked diurnal and seasonal fluctuations in the concentration of radon at a particular site (Harley, 1978, 1979; Fisenne, 1980). The total variation seems to be less than an order of magnitude, but it is apparent that extended measurements are required to evaluate outdoor exposure.

The indoor concentrations of radon are almost always higher than outdoor concentrations, largely because the vertical mixing cannot take place. In a few cases, the radon levels may be enhanced even more by elevated radium concentrations in building materials but, as a general rule, the major source of radon in single family dwellings is the soil directly under the building. A concrete cellar or concrete slab will, of course, retard the transfer from the soil to the air. Any cracks will be excellent channels for more rapid transfer (Landman, 1982). Cellars generally show the highest concentrations, both because they are nearest to the source and because they are poorly ventilated. Family living spaces will show concentrations which are almost entirely dependent on the degree of ventilation. For example, a rate of four air changes per hour will reduce indoor levels down to a value very close to outdoor air. As this ventilation rate is relatively uncommon, particularly in colder weather or in air-conditioned buildings, indoor concentrations will approach more closely to the equilibrium values obtained with no air changes at all. It should be noted that the desire to conserve energy has led to reduced intake of outdoor air and thus has tended to increase indoor radon concentrations. Two typical studies of indoor radon concentrations are those of George and Breslin (1978) and Cliff (1978).

3.4 Sources of Radon in the Global Atmosphere

The major source of radon in the atmosphere is the normal emanation from the radium-226 in the earth's surface. This and other sources are tabulated in Table 3.1. Table 3.2 lists the various assumptions required to calculate the contribution's of these various sources. A brief description of the sources of atmospheric radon follows.

3.4.1 *Natural Emanation from Soil and Water*

The global release of radon from soil can be approached in two different ways. One is to use the average emanation rate from the available measurements and assume that this holds for the land area of the earth. This procedure has been used in the tables of this section and gives a value of 2.4×10^9 Ci/y. A second approach is to assume that the earth's crust to one meter contains 1 pCi/g of radium-226. By taking the mean crustal density of 2.7, we can calculate that there are 4×10^8 curies of radium available. This would yield about 3×10^{10} curies of radon per year and a 10 percent emanation would produce a world release of 3×10^9 curies per year, which is in agreement with our other calculation. Both of these values are rough but do lead to world radon concentrations that fit the present data.

3.4.2 *Uranium Tailings*

Uranium ores, particularly in the United States, contain less than 0.5 percent uranium so that the mass of the residues produced on extraction is of the same order as that of the original ores. The chemical extraction of uranium leaves tailings sands, which contain

TABLE 3.1—*Sources of global atmospheric radon-222*

Source	Ci/y
Emanation from Soil	2×10^9
Ground Water (Potential)	5×10^8
Emanation from Oceans	3×10^7
Phosphate Residues	3×10^6
Uranium Tailings Piles	2×10^6
Coal Residues	2×10^4
Natural Gas	1×10^4
Coal Combustion	9×10^2
Human Exhalation	1×10^1

Note: All quantities have been rounded to one significant figure.

TABLE 3.2—*Values used in calculating sources of atmospheric radon*

Soil		
Emanation Rate	0.5 pCi/m²/s	(Guedalia, *et al*, 1970)
Land Area	1.5×10^{14} m²	
Radon Release	2.4×10^{9} Ci/y	
Oceans		
Emanation Rate	0.003 pCi/ m²/s	(Wilkening and Clements, 1975)
Ocean Area	3.6×10^{14} m²	
Radon Release	3.4×10^{7} Ci/y	
Natural Gas		
World Consumption	10^{15} L/y	
Concentration	10 pCi/L	(Barton *et al*, 1973)
Radon Release	10^{4} Ci/y	
Coal		
World Consumption	3×10^{9} tonnes/y	
Concentration	0.3 pCi/g	
Radon Release	900 Ci/y	
Ground Water (Potential)		
Evaporation and Transpiration	1×10^{14} m³/y	
Concentration	5000 pCi/L	(UNSCEAR, 1972)
Radon Release	5×10^{8} Ci/y	
Human Exhalation		
World Population	4×10^{9}	
Concentration of ^{226}Ra	50 pCi/person	
Radon Release	10 Ci/y	
Uranium Tailings Piles[a]		
Amount of Tailings	1.5×10^{9} tonnes	
Area Covered	12,000 hectares	
Emanation Rate	500 pCi/m²/s	(EPA, 1973)
Radon Release	2×10^{6}Ci/y	
Phosphate Residues[b]		
World Production-Rock	3.3×10^{8} tonnes/y	
Concentration of ^{226}Ra	40 pCi/g	
Emanation Fraction	0.1	
Radon Release	3×10^{6} Ci/y	
Coal Residues[b]		
World Coal Consumption	3×10^{9} tonnes/y	
^{226}Ra Concentration	0.3 pCi/g	
Emanation Fraction from Ash	0.01	
Radon Release	2×10^{4} Ci/y	

[a] All known and probable ore reserves extracted following EPA model mill pattern (EPA, 1973).
[b] Total production not known, so total residues assumed to represent 30 years accumulation at this rate.

some uranium and radium and the slimes, which are largely chemical precipitates designed to separate out impurities from uranium. These slimes contain high concentrations of radium but they are only a small fraction of the total residue weight. The radium concentration in the total tailings is several hundred pCi/g. At present, in the United States, there are about 50 sites containing uranium tailings with most of these being located in dry areas. About half of the sites are inactive.

About 20 percent of the radon formed in tailings is released and the emanation rates can be as high as 1000 pCi/m² per second. This can be reduced somewhat by covering the piles with inert soil but the factor is not large. One meter of a clay-texture soil reduced tailings emanation by a factor of 4 (Marple and Clements, 1978). While there is a requirement to reduce radon released from tailings in the United States, this is not true on a global basis.

While even the maximized global contribution of the tailing piles is not large, the concentrations described are a possible local problem. It is difficult to assess how much the high outdoor concentrations may contribute to indoor concentrations near the tailings piles and these latter may well control human exposure in the area.

3.4.3 *Phosphate Residues*

The phosphate extracted for fertilizers mostly occurs in calcium phosphate ores generally known as phosphate rock. In the United States, most of the phosphate rock is mined in Florida. The rock contains about 0.01 percent uranium-238 in equilibrium with its daughters (Guimond, 1977). Horton (1977) described the various types of waste piles associated with these residues.

The general processing is to react the crushed rock with sulfuric acid to produce phosphoric acid which is then further treated to produce various fertilizers. The major residues are sand tailings, low grade rock, and calcium sulfate (gypsum). The uranium follows the phosphoric acid path into the fertilizers while the radium generally remains with the calcium sulfate. The ores may contain more than 100 pCi of ^{226}Ra per gram with a common value of around 40. The radium concentrations in gypsum are in the range of 20 to 30 pCi/g which is lower than the ore due to the mass increase of hydrated calcium sulfate compared with the original calcium phosphate. The sand tailings contain undissolved radium plus fine crushed rock which is not processed and tend to contain less than 10 pCi/g of tailings.

Some gypsum is used for making cement or for preparing gypsum block or gypsum board, but a considerable amount goes onto the tailings piles. The major change in radon emanation caused by mining phosphate rock is bringing the material to the surface as compared with the normal soils in Florida, where the concentrations of radium are quite low.

3.4.4 *Coal*

The radium levels in coal appear to average about 0.3 pCi/g. The average coal in the United States contains about 13 percent ash which means that the radium carrying through to the ash would be concentrated to 2 pCi/g. At the time of combustion, the radon present will be released, but this is a relatively small amount compared with that which can be generated in the ash itself over a long time period.

The fly ash is somewhat enriched in radium, with the smaller sizes having the highest concentration (Coles *et al.*, 1978). The high temperature of combustion, however, tends to form rather glassy particles or other residues and these show a very low radon release, generally less than one percent of that formed in the ash. Beck *et al.* (1978) have calculated that the fly ash adds much less than one percent to the ambient radon levels. The ash residues are larger in quantity, but still do not contribute substantially to atmospheric radon. It does not appear that radon releases from coal are significant even on a local level.

3.4.5 *Ground Water*

The average concentration of radon in ground water is high compared to surface water. UNSCEAR (1972) indicated a mean of about 5000 pCi/liter while Hess *et al.* (1978) showed values up to 200,000 for granitic areas of Maine. Some fraction of this radon is released when the pressure is reduced, when the temperature is increased, or when the water is aerated. Another possible release mechanism is found when plants take up ground water. For example, Pearson (1967) found that the radon release from a cornfield was three times the release from bare soil.

It is not possible to give more than an upper bound to the possible global contributions from ground water. This has been done here by assuming the total evaporation and transpiration of water is from ground water and that all of the radon is released. There are estimates of local contributions from ground water supplies used in homes (Gesell and Prichard, 1978).

3.4.6. *Building Materials*

There are few data on the emanation of radon from building materials. Jonassen and McLaughlin (1978) found values of 0.001 to 0.25 pCi/m^2 per second for various concretes, with the emanation being proportional to the mass of material, rather than the exposed area. They found bricks, clapboard, fiberboard, and gypsum to release less than 10^{-4} pCi/m^2 per second. Fitzgerald and Sensintaffar (1977) estimated that gypsum board with 20 pCi ^{226}Ra/g would emanate at a rate of 0.1 pCi/m^2 per second.

The emanation rate from normal building materials thus is probably less than that from the average soil, and the rate is further reduced by the application of decorative coatings such as paint, wallpaper, and paneling. Although there are no definitive studies, it appears that the major source of indoor radon in most buildings is the soil beneath the structure. It should also be pointed out that many of the reported indoor concentrations have been measured in closed rooms. This is misleading to some degree as ventilation is the major controlling factor for indoor exposures.

3.5. Global Concentrations of Radon

The annual release of 2.4×10^9 curies of radon would lead directly to a global inventory of 3.6×10^7 curies, as there are no known sinks. This can be converted, in turn, to a surface concentration of 70 pCi/m^3 (Harley, 1973).

The average concentration is necessarily modified by the relative release rates over the continents and over the oceans. It is of interest to note that the fraction of land in the Northern Hemisphere is twice as great as in the Southern Hemisphere. Since the release from the oceans is very low, this means that the radon level in the Northern Hemisphere should be twice as great as that in the Southern Hemisphere. The calculated values for uniform distribution are 90 and 45 pCi/m^3, respectively, for each hemisphere. On the other hand, the levels would be about 250 pCi/m^3 for both hemispheres if we assume that all radon formed over the continents stays there. Both assumptions are obviously incorrect and the average concentrations fall between the two levels estimated. Turekian *et al.* (1977) have developed a model to describe the longitudinal distribution of radon in the Northern Hemisphere. The model fits the known data when a transit speed of 750 km/day is assumed.

3.6 Local Concentrations of Radon

Although the sources other than soil do not contribute significantly to global radon concentrations, they can have local effects. There are a number of measurements related to specific sources and these will be described here. The values do not necessarily represent anything other than the specific site and season of the measurements.

3.6.1 *Uranium Tailings*

Surveys of radon concentrations around uranium tailings piles have been reported by a number of authors. At a distance of 1 km from the source, enhancement of radon concentration is not measurable (Shearer and Sill, 1969); although high values have been reported near active mills (Eadie *et al.*, 1976). These measurements have been difficult, since background levels range from 500 to 1000 pCi/m^3 in these areas. This, in turn, derives from an elevated level of radium in the soils, which can be over 10 pCi/g soil.

3.6.2 *Uranium Mining Areas*

Uranium is mined both underground and in open pits. In the former case, radon formed underground is largely discharged to the atmosphere from ventilation shafts. In the second case, the radon disperses much the same as from the soil. The mine releases are not significant as global sources as compared with the tailings piles where many years of production residue is exposed.

Data on the mine releases are being collected by the Environmental Protection Agency (EPA), but a general evaluation is not yet available. One large open pit mine in New Mexico showed average outdoor downwind levels of over 1000 pCi/m^3 (Eadie *et al.*, 1979). This is a very large mine and other areas should show lower levels. Underground mines should be similar, as the radon release depends only on the amount of ore exposed.

4. Dosimetry of Inhaled Uranium Mine Aerosols

4.1 Introduction

Data on uranium mine atmospheres are essential before adequate exposure limits can be derived for radon and its daughter products. These data are required whether we derive standards via epidemiology or dosimetry. Formerly, these data often lacked sufficient detail (especially in older mines) to allow good estimates of exposure. Bronchial dosimetry is particularly dependent on good characterization of mine atmospheres, as many parameters go into the computation relating exposure to absorbed dose. With insufficient data, dosimetry as a method for exposure limitation is tenuous. With sufficient data, both physical and physiological, dosimetry ought to provide a firm basis for exposure limitation. The present discussion will focus on the mine environment and biological modeling, highlighting those areas felt to be of greatest concern for improving the dose assessment of individuals exposed to mine aerosols. Descriptions of the complex and highly variable radiation environment in uranium mines appear in many publications (USPHS, 1957; HASL, 1960; Holaday, 1964, 1969; JCAE, 1967; FRC, 1967; COO, 1968; Behounek, 1969; George and Hinchliffe, 1972; Cooper *et al.*, 1973; Hamrick and Walsh, 1974; NEA, 1976; ICRP, 1977a). No attempt will be made to expand upon these and other descriptions; they will be used merely to draw attention to the parameters that influence the amount of energy absorbed by lung tissue.

While other sources of radiation exposure will not be considered in detail here, it should be noted that miners are also exposed to external gamma radiation and to inhalation of uranium ore dust.

External gamma radiation in domestic uranium mines seldom exceeds 2.5 mR per hour, and the average is only a fraction of this (FRC, 1967). Excluding scattered radiation, the average gamma ray energy is about 1 MeV. The exposure rate varies with location in a mine in a manner dependent on the radium content of the surrounding rock. In very rich ores (about 20–30 percent U_3O_8 by weight), exposure rates on the order of 100 mR per hour have been observed (ICRP, 1977a). Because beta radiation is more easily shielded, it is of less importance as an external hazard, even though its radiation dose in air may be higher than the gamma dose by a factor of 10 (FRC, 1967).

Ore dusts, containing members of the uranium and thorium decay series, are dispersed into the air from mining operations and become sites of attachment for the airborne radon daughters. Condensation nuclei brought into the mines from the outside during ventilation, and other aerosols (such as diesel smoke) associated with mining operations, compete with ore dust as carriers of the radon daughters. The short-lived radon daughters are, by now, universally recognized as the principal radiation hazard from exposure to uranium mine aerosols, although in earlier times radon gas was thought to be responsible for the radiation-induced morbidity and mortality among uranium miners. Unlike some foreign mines, U.S. mines are relatively free of natural thorium (FRC, 1967). Where natural thorium is abundant, an alpha dose to bronchial epithelium can occur from thoron daughters. Thoron daughter dose must also be considered in the occupational exposure limitations but is not included in this report.

4.2 Origin of Radon and Radon Daughters in Mines

Radon gas is produced by decay of ^{226}Ra. Radon diffuses into mine air from rock surfaces and from ground water. From there it is carried in ventilation currents, decaying to the short-lived solid decay products: RaA (^{218}Po), RaB (^{214}Pb), RaC (^{214}Bi) and RaC′ (^{214}Po). Radon concentration increases in proportion to emanation rates and residence time of the air in the mine passages. The short-lived daughters build up rapidly with time and can approach equilibrium with the radon. Once produced, the daughter products exist in atomic form for a period of time dependent upon the availability of aerosol surfaces for attachment. The ultimate radioactive aerosol contains attached and unattached daughter products. The percent unattachment is invariably largest for RaA. Because carrier aerosols and other surfaces collect these highy diffusible "free ions" or "ultrafine particles", the subsequent decay of RaA to RaB and RaC-C′ results in these later daughters being attached.

From a radiation protection viewpoint, the daughter products which can deposit on the bronchial tree deliver the highest alpha dose. Thus, even though an external radiation hazard is associated with work in a uranium mine environment, the principal hazard stems from the internal exposure to the short-lived alpha-emitting daughters of radon formed in the mine air.

Since radon gas is chemically inert, it is capable of considerable migration from its point of origin and is present everywhere in air to some degree. Because of varying ventilation and filtration, and the presence of differing carrier aerosols, the characteristics of the exposing atmosphere may be appreciably different for the mine (whether uranium or other types of mines) than for the home, the outdoors, or other work environments. Consequently, it is not surprising that the radiation dose varies for these different exposure conditions.

Furthermore, each individual mine environment has its own complex and changing exposure conditions. Since dosimetry must treat average or mean exposure conditions, it is imperative to account for any temporal changes in the "average" conditions for various phases of development of the uranium mining industry. Exposure conditions must be well-defined and must truly represent the norm before dosimetry can be applied with any degree of confidence to a particular exposure situation in time or place.

4.3 Calculation of Radiation Dose

A large number of models have been developed over the years to calculate the radiation dose to the lung as a whole or to portions of the lung. Table 4.1, covering the time span from 1940 to 1981, presents a summary of much of this work (Cross *et al.* (1974) updated for this report. The tabulated data show not only the evolution in thinking and the increased sophistication in modeling, but the variation in the dose conversion factor (rad/WLM) and in derived yearly limits for exposure to both radon gas and radon daughters. A general trend seems to be a gradual lowering of the dose conversion factor to unity or less as the unattached fraction decreases and the activity becomes attached to intermediate-sized diesel smoke particles.

4.3.1 *Radon Gas*

It should be stressed that under usual exposure conditions, the alpha-emitting short-lived daughters of radon, rather than the radon gas, are the recognized principal hazard. However, it is of interest to derive an upper limit on exposure to radon gas alone in the absence of the daughters. The MPC derived by ICRP (1959) under these conditions is 3000 pCi/L to produce a dose of 15 rem/year to the lung ($Q = 10$). Others (Cross *et al.*, 1974) have calculated values close to

TABLE 4.1—*Summary of published dose calculations and recommendations for occupational exposure to radon and radon daughters*

Investigator	Remarks	Critical Tissue	rad/WLM	^{222}Rn[a] (pCi/L)		WLM/y[a]
Evans and Goodman (1940)	10 pCi/L Rn recommended; presumably for plant, laboratory or office air rather than mine air	—	—	10	→	1.2
Evans (1940)[b]	$(MPC)_a$ calculation for Rn + daughters of its decay (19.2 MeV effective α-ray energy)	Bronchi of right lung (9.8 g)	—	3400		—
NBS Advisory Committee (1941)	10 pCi/L Rn recommended	—	—	10	→	1.2
British X-Ray and Radium Protection Committee (1943)	1000 pCi/L Rn recommended	—	—	1000	→	12
Mitchell (1945)	$(MPC)_a$ calculation for Rn + daughters of its decay (6.95 MeV effective α-ray energy)	Larger bronchi (1 cm diameter)	—	970		—
British X-Ray and Radium Protection Committee (1948)	50 pCi/L Rn recommended	—	—	50	→	6.0
Evans (1950)	$(MPC)_a$ calculation for Rn + daughters of its decay (19.2 MeV effective α-ray energy)	Larger bronchi (1 cm diameter)	—	360		—
Bale (1951)	$(MPC)_a$ calculation for Rn + daughters of its decay (19.2 MeV effective α-ray energy)	Lung (2.75 L)	—	6600		—
	$(MPC)_a$ calculation for Rn + daughters of its decay (19.2 MeV effective α-ray energy)	Bronchi (19.6 g)	—	3500		—

TABLE 4.1—*(Continued)*

Investigator	Remarks	Critical Tissue	rad/WLM	^{222}Rn[a] (pCi/L)		WLM/y[a]
	Calculation for RaA 100%, RaB 50%, and RaC 50% equilibrium, and 40% bronchial retention for 1 hr	Bronchi (19.6 g)	7.5	—		0.20
AEC Biophysics of Radium and Radon Symposium (1951)	Recommended by those present that the value originally chosen in 1941 for occupational exposure be retained	—	—	10	→	1.2
Morgan (1951, revised 1954)	Calculation for RaA 100%, RaB 50%, and RaC 50% equilibrium, and 12% bronchial retention	Bronchi (20 g)	2.6	—		0.58
ASA (1952)	10 pCi/L Rn recommended	—	—	10	→	1.2
NBS (1952)	10 pCi/L Rn + equilibrium daughters recommended. Based on recommendations of NBS and the ASA in 1952 and calculations of K. Z. Morgan in 1951	—	—	10	→	1.2
Tripartite Harriman Conference (TRI-C, (1953)	100 pCi/L Rn + equilibrium daughters recommended. This choice was presumably a balancing of risks versus benefits.	—	—	100	→	12
Shapiro (1954)	Calculation for RaA 100%, RaB 100%, and RaC 100% equilibrium; and 25% lung retention (typical of measurements in human subjects breathing "normal atmospheric dust")	Lung (1000 g)	0.12 (corrected to 20.8 L/min breathing rate)	—		13.0

	Calculation for equilibrium daughters on typical atmospheric condensation nuclei using Findeisen's Lung Model	Tertiary bronchioles (≃Weibel Generations 4, 5 and 6)	0.31 (corrected to 20.8 L/min breathing rate)	—		4.8
	$(MPC)_a$ calculation for Rn + daughters of its decay (19.2 MeV effective α-ray energy)	Lung (1000 g)	0.002 equivalent (assuming radon daughter equilibrium)	5600		—
	$(MPC)_a$ calculation for Rn + daughters of its decay (19.2 MeV effective α-ray energy)	Trachea	—	210		—
Bale and Shapiro (1956)	Estimated on the basis of experiments that 25 pCi/L Rn + equilibrium daughters delivers 0.3 rem/wk to the bronchi	Bronchi	0.50	—		3.0
Seven State Uranium Mining Conference (Holaday, 1955)	Recommended by those present that radon concentrations be disregarded as a measure of control and that the "working level" of 1.3×10^5 MeV/L of alphas from the short-lived daughters be adopted	—	—	—		12
ICRP (1955)	100 pCi/L Rn + equilibrium daughters intended recommendation. Choice was based on discussions at Tripartite Conference	—	—	100	→	12
Chamberlain and Dyson (1956)	Calculation assuming only the RaA alphas of uncombined RaA atoms (abundance = 10% of the equilibrium value for RaA) contributed to the dose	Trachea plus main bronchi	0.38	—		3.9

TABLE 4.1—*(Continued)*

Investigator	Remarks	Critical Tissue	rad/WLM	^{222}Rn[a] (pCi/L)	WLM/y[a]
Hultqvist (1956)[c]	Calculation of 150 mrem/yr to "lung tissue" from 0.3 pCi/L Rn + equilibrium daughters. This value is cited in NAS/NRC Publ. 848, "Effects of Inhaled Radioactive Particles," and is said to represent the dose rate to the bronchi	Bronchi	0.42	—	3.0
U.S. Public Health Service (USPHS 1957)	100 pCi/L Rn + daughters → 1WL recommended	—	—	—	13
	$(MPC)_a$ calculation for Rn + daughters of its decay (19.2 MeV effective α-ray energy)	Lung (1000 g)	—	6300	—
	$(MPC)_a$ calculation for Rn + daughters of its decay (19.2 MeV effective α-ray energy)	Bronchi (20 g)	—	3300	—
	$(MPC)_a$ calculation for Rn + daughters of its decay (19.2 MeV effective α-ray energy)	Bronchi (1 cm diameter)	—	290	—
	Calculation for RaA 100%, RaB 50%, and RaC 50% equilibrium, and 25% lung retention	Lung (1000 g)	0.11 (corrected to 20.8 L/min breathing rate)	—	14.0
	Calculation for RaA 100%, RaB 50%, and RaC 50% equilibrium, and 12-1/2% retention in the bronchi	Bronchi (20 g)	1.4 (corrected to 20.8 L/min breathing rate)	—	1.1
ICRP (1959)	30 pCi/L Rn recommended assuming RaA 100% equilibrium and 10% unattached RaA ions	—	—	30 →	3.6

	100 pCi/L Rn recommended, assuming RaA 100% equilibrium and 3% unattached RaA ions	—	—	100	→	12
NBS (1959)	30 pCi/L Rn recommended, based on same assumptions as ICRP (1959)	—	—	30	→	3.6
ASA (1960)	Established minimum requirements for protection of uranium miners: Radon daughter standard—1 WL Cease work level—10 WL	—	—	—		12
AEC (1960)	100 pCi/L Rn maximum permissible concentrations (MPC)[a]	—	—	100	→	12
Governors' Conference (USPHS, 1961)	Those present reaffirmed the 1 WL standard chosen at the 1955 Seven State Conference. An increased lung cancer risk was first documented at this conference.	—	—	—		12
IAEA (1962)[d]	"International Standard" for uranium mining provisionally adopted at 3 WL, presumably because the 1959 ICRP value was impractical to apply.	—	—	—		36
Stewart and Simpson (1964)	After a comprehensive review of all information available to them, representing the accumulated thought, experiment and experience of over 20 years, confirmed the value and practical usefulness of the 1940 Evans and Goodman value.	—	—	10	→	1.2

TABLE 4.1—*(Continued)*

Investigator	Remarks	Critical Tissue	rad/WLM	^{222}Rn[a] (pCi/L)	WLM/y[a]
Altshuler, et al. (1964)	Calculation using the respiratory tract model of Findeisen (as improved by Landahl) for a "typical" mine atmosphere of 200 pCi/L of daughters with 4.7% of the activity as ions, 55.5% on condensation nuclei and 40% on particles >0.1 μm diameter. As very little of the dose was found to be contributed by the condensation nuclei, it is assumed their calculations also pertain to a breathing rate of 20.8 L/min.	Basal cells in the segmental bronchi (≃Weibel Generations 4, 5 and 6) at 36 μm depth	3.5 (mouth breathing)	—	0.43
			1.9 (nose breathing)	—	0.80
Jacobi (1964)	Calculation of mean α-dose rate using a conceptually similar model to that of Altshuler, *et al.* (1964), but for all attached activity on condensation nuclei. The differences in these two models have been detailed in various reviews. A consensus is that the agreement in the region of the lower bronchi is highly fortuitous and disappears if the same depth to the target tissue and the same values for the alpha particle ranges are used in both.	Sec.—quart. bronchi (≃Weibel Generations 2–9)	3.1 (mouth breathing)	—	0.49

Haque and Collinson (1967)	Refined the previous dose calculations using the diffusion theory of Gormley-Kennedy, the Weibel lung model and a calculational technique for determining the distribution of the doses with depth into the bronchial epithelium. Their values were generally highest in the segmental bronchi but varied widely with the reference atmosphere. The exampled calculation is for sites containing luminous paints where RaA/RaB/RaC = 1/0.86/0.76 and free ion fractions = 0.35/0.58/0.077 with the remainder of the activity on condensation nuclei. Nose penetration factor for both ions and nuclei was assumed to be 0.75 (a crude assumption when the majority of the dose is from the free ions).	Segmental bronchi (Weibel Generation 5) at 35 μm depth	11.0 (mouth breathing)	—	0.14
			8.0 (nose breathing)	—	0.19
FRC Staff Report (1967)	Based mainly on a comprehensive review of the previous dosimetry models by H. M. Parker, the FRC staff concluded with reservations that the working level month (WLM) corresponded to 2.8 rad. The differences among the various	—	2.8	—	0.54

TABLE 4.1—*(Continued)*

Investigator	Remarks	Critical Tissue	rad/WLM	^{222}Rn[a] (pCi/L)	WLM/y[a]
	dosimetry models and radiobiological parameters were considered too large for use in estimating the risk of radiation-induced carcinogenesis.				
U.S. Dept. of Labor (FR, 1967a)	The DOL with enforcement authority under the Walsh-Healy Public Contracts Act put into effect "Radiation Standards for Uranium Mining" (41CFR50-204.321) establishing 3.6 WLM/yr, changed later in 1968 to 4 WLM/yr.	—	—	—	4.0
FRC (FR, 1967b)	The FRC in its "Radiation Protection Guide for Federal Agencies" adoptd the 1 WL standard for uranium mining after initially considering an exposure guidance of 36, 12 and 4 WLM/y. Presumably 12 WLM/y contained the best balance of benefit-cost-risk in the light of available knowledge.	—	—	—	12
Holleman (1968)	Calculation for 35-μm depth using energy-expended values similar to Haque and Collinson, but instead of determin-	Bronchi at 36-μm depth	0.8 ± 0.2 (avg. breathing pattern during work)	—	1.9

	ing the equilibrium areal activities from the deposition rate and the mucus movement, Holleman used the measured deposition in working miners and assumed it was distributed uniformly throughout the 6000-cm^2 area of the TB tree. The relative abundance of daughters measured in a Colorado hardrock mine was 1/0.53/0.35 and the free ion values determined after the method of Craft, et al. were about 10% for RaA and <3% for RaB and RaC.				
FRC (FR, 1969)	Reaffirmed its 1967 recommendations as interim guidance but declared an intent to reduce the level to 4 WLM/yr in 1971 pending further review.	—	—	—	12
U.S. Dept. of Interior (FR, 1969b)	The DOI published its proposed rules and regulations pursuant to the Federal Metal and Nonmetallic Mine Safety Act requiring exposure not to exceed 12 WLM/yr.	—	—	—	12
Nelson, *et al.* (1969)	Calculations after the method of Haque and Collinson for ions and condensation nuclei, and Landahl for deposition of the larger particles. Nelson used	Segmental bronchi (Weibel generation 4) at 35μm depth	14 (mouth breathing)	—	0.11

TABLE 4.1—*(Continued)*

Investigator	Remarks	Critical Tissue	rad/WLM	^{222}Rn[a] (pCi/L)	WLM/y[a]
	Holleman's radon daughter ratios; assumed unattached ion fractions to be 0.10/0.01/0.01; placed 10% of the total activity on 6-μm diameter particles and the remainder of the activity on condensation nuclei describable by a mean diffusion coefficient of 1.3×10^{-5} cm^2/s. Because of uncertainties in application of the Gormley-Kennedy diffusion equations due to likely turbulence and "end effects" in the proximal airways, a factor of two increase in deposition was assumed to apply to ions and nuclei throughout the model lung. An interim cancer-related dose was selected to be 10 rad/WLM, probably accurate to within a factor of three.		6 (nose breathing) 10 (equal combination)	— —	0.25 0.15
Walsh (1970)	Calculation using an altered Altshuler reference atmosphere to reflect the measurements of Holleman, who reported no significant activity on particles >0.1 μm in diameter. Walsh retained the free ion activities of Altshuler but placed the remainder of the activity on condensation nuclei. Uniform dep-	Bronchi Tertiary bronchioles (Weibel Generations 4, 5 and 6) at 36 μm depth	1.6 0.5	—	0.94 3.0

	osition of the activity on the surface of the T-B tree was assumed, as well as no translocation. Assuming all the alphas reach the tissues of interest, he calculated an average T-B dose of 1.6 rad/WLM. His 36-μm depth dose in the tertiary bronchioles is about 0.5 rad/WLM with a stated maximum of 1 rad/WLM.				
US Dept. of Interior (FR 1971)	Airborne concentrations not to exceed 1 WL	—	—	—	4.0
USEPA (FR 1971)	—	—	—	—	4.0
USEPA (1971)	Presidential Reorganization Plan of 1970 transferring the function of the FRC to EPA. After a review and analysis of material prepared by various advisory groups, the EPA administrator reaffirmed the previous FRC guidance of 4 WLM/yr to be effective July 1, 1971.	—	—	—	4.0
NIOSH/NIEHS (1971)	In a joint monograph, "Radon Daughter Exposure and Respiratory Cancer Quantitative and Temporal Aspects," concluded that 1 WLM is equivalent to about 2 rad averaged over the tracheobronchial epithelium and that the quality factor for alpha particles is probably about 3.	Bronchi	2.0	—	0.75
U.S. Dept. of Interior (CFR, 1972)	Under mandatory rule 57.5-42 declared its intention to ob-	—	—	—	4.0

TABLE 4.1—*(Continued)*

Investigator	Remarks	Critical Tissue	rad/WLM	^{222}Rn[a] (pCi/L)	WLM/y[a]
	serve the recommendations of EPA in keeping with their previously stated intentions of following the recommendations of the appropriate authority issuing guidance.				
Jacobi (1972)	Calculation of ranges of dose rates employing the 1966 ICRP lung model, which allows activity transport other than ciliary-mucus, and the model Jacobi developed for determining inhaled potential α-energy as a function of aerosol concentration, ventilation rate and mine wall-deposition rate.	Bronchi (100 g) (Presumably Weibel Generations 0–16)	0.4–1.0 (nose breathing & low aerosol concentration)	—	1.5–3.8
			0.07–0.12 (nose breathing & high aerosol concentration)	—	12–21
Harley and Pasternack (1972)	Calculations employing the Gormley-Kennedy diffusion equations and the Weibel lung model, but with mucus clearance times determined by the method of Altshuler, et al., rather than the times employed by Haque and Collinson. They assumed uniform mixing of the activity within the mucus and employed the latest mine aerosol measurements of George (HASL-TM-70-7, 1970), which indicated	Segmental bronchi (22 μm in tissue or about 36 μm from mucus surface)	0.23 (nose breathing & RaA/RaB/RaC = 1.0/1.0/1.0)	—	6.5
			0.36 (nose breathing & RaA/RaB/RaC = 1.0/0.48/0.34)	—	4.2
			1.1 (nose breathing & RaA/RaB/RaC = 1.0/0/0)	—	1.4

	that for mines making extensive use of diesel equipment, the RaA free ion abundance is <5% and the remainder of the activity is attached to particles with a median size of 0.2 to 0.4-μm diameter.				
Jacobi and Eisfeld (1980)	Weibel model, correction for turbulence in generations 0–5, desorption of particles to blood.	Basal cells, whole lung, kidney, liver, spleen, stomach	8+70 fpot[e]		3.5–6
Harley (1980)	Dosimetry based upon a five-lobed anatomically correct human male lung. Correction for convective diffusion (turbulence) in Generations 0–5.	Shallow basal cells in bronchial epithelium (22 μm below epithelial surface).	0.58 (average value of basal cell dose to Generation 4 in the lobes).		3
ICRP (1981)	Annual limit based upon a judgment of dosimetric and epidemiologic evidence.	Separate calculation for whole bronchial tree, whole lung and basal cells in bronchial epithelium	8+70 fpot (basal cells) 6.4+40f pot (whole bronchial tree 45 grams)	4000[f]	4.8

[a] Calculated on the basis of an MPD of 0.3 rem/40-hr wk or 15 rem/yr. RBE_{α} = 10 and breathing rate = 20.8 L/min. The WLM is an exposure for 170 hr to any combination of the short-lived radon daughters (RaA, RaB and RaC-C′) in 1 liter of air that will result in the ultimate emission of 1.3×10^5 MeV of potential alpha energy. WLM/y values are derived from the radon concentrations, assuming radioactive equilibrium of the short-lived daughters.

[b] Stewart and Simpson, 1964. Evans work appears in Jackson (1940) unpublished.

[c] Spiers, 1956

[d] JCAE, 1967

[e] fpot = the total potential alpha energy of the unattached daughters.

[f] Calculated assuming that all short-lived daughters formed from radon in the lung decay there and applying a weighting factor Wt = 0.12 for the lung.

this (5000 pCi/L) assuming a 3.3-liter lung air volume and taking into account the small additional absorption of radon in the lung tissue. Selecting a functional residual lung capacity of 3.5 liters and an average tidal air volume of 0.5 L, a recalculated upper limit based upon whole lung dose from exposure to radon gas alone is about 4000 pCi/L. ICRP (1981) derive a limit of 4000 pCi/L using $Q = 20$. Thus, values range from about 3000 pCi/L to 5000 pCi/L. Radon concentrations are of practical importance, although perhaps only where miners must resort to respirators for the removal of airborne daughter products.

4.4 Factors Influencing Radon Daughter Radiation Dose

A detailed discussion of the variability in dose to exposure ratios introduced by the various modeling parameters is given in Section 6. Some of the factors specific to underground mines are described in this section.

No attempt will be made to trace the exposure conditions that previously existed, or how they may have changed with time in the mines. Although it has been suggested that the aerosol characteristics have remained essentially unchanged with time (Section 6.2), this is difficult to accept in the light of our previous speculations (Section 4.2) concerning carrier aerosols. Over the years, mechanical ventilation has replaced natural ventilation and machinery has replaced hand labor, considerably altering conditions. The calculations of Jacobi (1972), Harley and Pasternack (1972), and Cross *et al.* (1974) demonstrate the influence on dose of changes in aerosol concentration which affects the unattached fraction, particle size, and degree of daughter disequilibrium. All of these parameters influence the dose conversion factor and are necessary in any discussion on the adequacy of the WLM unit for assessment of the inhalation risk.

4.4.1 *Percent Unattachment, Particle Size, Daughter Disequilibrium*

Regardless of what the percent unattachment may have been in the past, there is general agreement that the average value now lies in a range of about 2 to 4 percent (George and Hinchliffe, 1972; Sciocchetti, 1976; Pradel, 1973). Because of the relationship between inert aerosol concentration and percent unattachment, it is unlikely that older mines differed by large factors (see Section 6.2). The diffusion coefficient of unattached daughter products has been traditionally accepted as 0.054 cm^2/s, but there is increasing evidence to suggest a value of 0.015 cm^2/s (Pradel, 1973; ICRP, 1977a) or lower [0.005 cm^2/s (Harley and Pasternack, 1981)] for other than freshly formed RaA. The latter values, although not changing overall lung deposition probability, influence the estimate of the magnitude and region of highest dose in the tracheobronchial tree, that part of the lung pertinent in human lung cancer induction.

Activity median aerodynamic diameters are found to range from 0.2 to 0.4 μm, with a geometric standard deviation of about 2, but they may be lower in some mines (George *et al.*, 1970). Particle concentration values range from $10^6/cm^3$ after blasting to $10^5/cm^3$ in working areas, $10^4/cm^3$ in drifts, and between 10^3 and $10^4/cm^3$ in nonworking areas of the mine or outside of the mine (George and Hinchliffe, 1972; Pradel, 1973). The particle concentration is of interest since it influences the degree of daughter attachment and possibly, lung clearance mechanisms. The degree of daughter equilibrium ranges from near 100 percent to the extreme case of only RaA being present. This can occur only under conditions of high local filtration, and the average is 1/0.53/0.35 for RaA, RaB, and RaC, respectively, for a Colorado hardrock mine (COO, 1968). This is not too different from Altshuler's reference atmosphere (1/0.66/0.47). The mean value for drifts in French mines is (1/0.60/0.30) (Pradel, 1973). All of these ratios would yield comparable dose values.

All lung dosimetry models except that of Altshuler *et al.* (1964), including that of the ICRP (ICRP, 1978), ignore the spectra of particle sizes and assume that median diameters adequately define deposition probabilities. It should be noted that when diffusion equations are used to calculate deposition probabilities, the spectra could be used. For practical reasons, however, median sizes are employed to reduce the computational effort. The means of other physical parameters are also used in dose calculations. It is highly unlikely, however, that greater refinement in the physical description of the aerosol beyond mean values for degree of disequilibrium, percent attachment, activity median diameter, and activity concentration will yield commensurately greater accuracy in the dose calculations for miners. The physiologic parameters that influence the amount of energy absorbed in lung tissue appear, at this time, to be the limiting factor in radiation dosimetry.

4.4.2. *Clearance Rates and Target Tissue Depth*

To model radon daughter activity in the lung, standard or reference physiologic parameters are used. As these are also based on mean values for populations, the dose assessment for an individual might contain large errors. Two sources of variability in calculated dose attributable to physiological factors are mucus clearance rate and target cell depth.

Measurements on humans and animals have indicated both fast and slow clearance rates for radon decay products in the lung (Palmer *et al.*, 1964; IAEA, 1973; Jacobi, 1976; ICRP, 1966). Disease states of the lung influence this parameter and may also alter the thickness of tracheobronchial mucus and epithelium, thereby affecting the amount of alpha energy that can be absorbed by basal cells, which are the generally accepted target at risk.

The depth of the basal cell in bronchial epithelium is important for calculation of dose to man. Their minimal depth is on the order of 35 μm beneath the surface of the mucus (FRC, 1967) or ~22 μm below the epithelial surface. Fortuitously, there is little numerical difference in calculated dose between a model that uniformly loads the mucus and serous layers (combined thickness is assumed to be 15 μm) with activity, and calculates the dose at a depth of 22 μm below the epithelial surface which places all of the activity on the surface of the mucus and calculate dose at 35 μm below this surface. This is a consequence of the fact that a large fraction of the dose comes from the RaC′ alpha, and its energy is sufficiently high that there is little difference in a "thin" or "thick" source calculation. What is fortuitous about the 35 μm minimal depth for the basal cell is that the dose, calculated as an average over the range of the RaC′ alpha (70 μm), yields agreement to within about 30 percent of that calculated at 35 μm with more complicated models.

4.4.3 *Dose Calculation Model*

There are other physiological parameters that influence the tracheobronchial dose, such as mouth vs. nose breathing, tidal volume, etc., but, again, the major difference in the calculated doses to the tracheobronchial region arises from the different assumptions made concerning particle size, mucociliary or other biological clearance rates and target cell depth (see Section 6 for detailed calculations). Harley and Pasternack (1972) calculated clearance for the Weibel lung model (see Table 5.3, Weibel, 1963) based on values adopted by Altshuler *et al.* (1964). They calculated doses to the segmental bronchi similar to those of Altshuler and conclude that, within the framework of any model, the mucus transit times for radon daughters are not critical because the dose would only be increased by 50 percent if there were no clearance. The modified model of Haque and Collinson (1967) [using their corrected equivalent number of alpha particles per square micron which deposit their entire original kinetic energy over a 1 μm segment of track (Haque, 1967)] assumes mucus transit in Weibel Generation 1 and Generations beyond 3 much different than those adopted by Harley and Pasternack (1972) and Jacobi (1964).

Table 4.2 presents a detailed breakdown of calculations using two tracheobronchial models for nose breathing (assuming 40 percent penetration for the unattached activity and 98.7 percent penetration for activity attached to 0.3 μm diameter particles). In general, the dose differences appear to parallel the mucus transit times (the longer the activity resides in a region, the greater is the dose to that region). The Haque and Collinson model uses a breathing rate of 20.8 L/min, whereas the Harley and Pasternack model utilizes 15 L/min. Table 4.2 indicates that the average dose ratios for the two models vary between roughly 1.5 and 3.0, in a manner somewhat predictable by the mucus transit time ratios, but the dose ratio in any one generation

TABLE 4.2—*Comparison of regional dose and mucociliary clearance ratios*[b]

Weibel Generation	Harley-Pasternack Model rad/y	Haque-Collinson Model rad/y	Dose Ratio	Mucus Transit Time Ratio[c]
0	0.68	1.1	1.6	1.2
1	0.95	0.86	0.91	0.63
2	2.2	2.7	1.2	0.95
3	1.8	2.1	1.2	1.0
4	2.8	5.1	1.8	1.8
5	2.0	3.3	1.7	1.8
6	1.3	1.6	1.2	1.8
7	2.3	5.5	2.4	$\sim\infty$
8	1.4	5.2	3.7	$\sim\infty$
9	0.71	4.8	6.8	$\sim\infty$
10	1.8	4.2	2.3	$\sim\infty$
11	1.7	3.8	2.2	$\sim\infty$
12	1.5	3.3	2.2	$\sim\infty$
13	1.9	3.1	1.6	$\sim\infty$
14	1.3	2.9	2.2	$\sim\infty$
15	1.1	2.6	2.4	$\sim\infty$
16	0.75	2.4	3.2	$\sim\infty$
4–6[a]	1.9	3.0	1.6	
2–9	1.5	4.4	2.9	
0–16	1.2	2.9	2.4	

[a] Regional doses are averaged by surface area.

[b] Atmosphere Characteristics; 100 pCi/L each of RaA, RaB, RaC, 4 percent unattached RaA.

[c] The Haque-Collinson model assumes no clearance in generations 7 to 16.

may vary even more. This variability may or may not be significant, according to whether high local doses assume significance over regional averages. According to NCRP Report No. 39 (NCRP, 1971), there is sufficient tissue mass or surface area in any *one* generation for suitable dose averaging.

In view of the apparent influence of mucus velocity on the calculated regional doses in the tracheobronchial tree, it is prudent to review this parameter, selecting concensus averages for "normal" individuals noting changes for disease states and ciliastatic substances such as tobacco smoke. Table 4.3 presents a summary of the mucus transit times used in some of the more sophisticated dosimetry models. It should be noted that Jacobi's values would result in lower doses beyond Generation 3 than either of the other models. The transit times are such that biological removal rates are of far greater import to the tracheobronchial doses. This is in sharp contrast to the other two models which, because of their longer retention times, would produce larger doses in the region of importance to carcinogenesis (Generations 2–9) (Turner *et al.*, 1977). The ICRP lung model for soluble aerosols (retention Class D for radon daughters) assumes that 95 percent of the activity deposited in the tracheobronchial tree is removed to blood

TABLE 4.3—*Mucus transit times for three lung dose models*

Generation	Transit Time in Minutes		
	Harley-Pasternack[a]	Haque-Collinson[b]	Jacobi[c,d]
0	8	9.6	8
1	6	3.8	3.2
2	8	7.6	9.5
3	3	3.04	3.8
4	14	25.4	6.4
5	12	21.4	5.4
6	10	18.0	4.5
7	32	$\sim\infty$	3.8
8	27	$\sim\infty$	3.2
9	23	$\sim\infty$	2.7
10	445	$\sim\infty$	9.2
11	378	$\sim\infty$	7.8
12	320	$\sim\infty$	6.6
13	261	$\sim\infty$	5.4
14	223	$\sim\infty$	4.6
15	194	$\sim\infty$	4.0
16	160	$\sim\infty$	3.3

[a] Harley and Pasternack, 1972.
[b] Haque and Collinson, 1967.
[c] Jacobi, 1964.
[d] IAEA, 1973.

with a clearance half-time of 14.4 minutes, while the remaining 5 percent is removed by mucociliary transport with a half-time of 288 minutes (ICRP, 1978). Beginning in 1981, the ICRP (ICRP, 1981) calculated the dose in two ways. As one approach, the above lung model was used and the dose to the total 45-gram mass of the bronchial tree was calculated. This alpha dose is similar numerically to their calculation using either of two bronchial dose models based on the Weibel (1963) and Yeh-Schum (1980) morphometry (Jacobi and Eisfeld, 1980; James *et al.*, 1980). However, only average dose over the bronchial tree is given (generation 2–15) in any of the bronchial models, not the dose to a specific bronchial airway. These average absorbed alpha dose values are shown in Table 4.1.

The whole question of clearance parameters must be brought into sharper focus. It can only be assumed that the faster clearance times selected by Jacobi and ICRP reflect experiments, such as those involving attachment to room aerosols, where the removal rates were rapid. It must be determined, however, whether these faster rates adequately describe attachment to mine aerosols and the somewhat impaired conditions of stressed lungs. In summary, choice of clearance rates needs resolution.

Desrosiers (1977) has speculated that some of the dose differences among models are due to errors in calculating energy expended rather than in mucus transit times, as we have shown. Although his treatment appears convincing, there is inconsistent evidence that Haque (1967a, b) and those employing Haque's energy-extended model have not properly accounted for the energy laid down in microscopic volumes. The uncertainties noted by Nelson *et al.* (1974) when employing Haque's model, were subsequently resolved by Haque (personal communication to I.C. Nelson, January 2, 1970), who found a systematic error in the Haque treatment that progressively lowered areal equilibrium activities from the true values as the trachea was approached. The difference between Haque and Nelson *et al.* that were noted by Desrosiers were of a magnitude that happened to fit Desrosiers speculations as to why the differences existed. It is concluded, therefore, that the calculational method of Haque and Collinson (1967) (using an equivalent number of alpha particles, N_{eq}, which deliver the full alpha energy at depth), when employing the updated N_{eq} values (Haque, 1976b), provides an adequate description of tracheobronchial dose (see Section 6), but that differences in dose among models are still attributable to differences in the selection of the various parameters which are necessary for the total dose model. These are described in detail in Section 6.

4.5 Other Influences on Dose

Some thought might also be given to weighting the dosimetry by the number of basal cells in each generation. This idea, although not new, represents an interesting point of departure from the ICRP methodology of averaging the dose over the tracheobronchial tree (ICRP, 1981). It is estimated that when the depth of basal cells is taken into account, using the values of Gastineau (1969), the dose to the basal cells is about 40 percent of that to the entire tissue (all cells) of the tracheobronchial tree. A counterbalancing factor (increased deposition at bifurcations) could also be applied. Chamberlain and Dyson (1956) and Martin and Jacobi (1972) indicate that, for ions and particles, the activity per unit area at the junction of the trachea and main bronchi is perhaps two or three times that of nearby surfaces. Whether this is true at other bifurcations is uncertain, as is the thickness of epithelium at these sites. If, for example, basal cells lie deeper at bifurcations, the enhanced deposition at these sites is of less consequence.

Another related factor affecting dosimetry is the correction for nonlaminar flow in the larger bronchi. Martin and Jacobi (1972) present ratios of observed to expected deposition probabilities, based on calculated depositions and experiments with a plastic model of the upper bronchial tree. Employing their correction factors, it is estimated that the dose to the segmental bronchi (generation 4) is increased approximately 30 percent, whereas the dose averaged over generations 2–9, or over the whole tracheobronchial tree (generations 0–16) is essentially unaltered.

Finally, the choice of quality factor (Q) or relative biological effectiveness (RBE), specifically for alphas in the lung, needs addressing in view of the current ICRP choice of 20 for Q (ICRP, 1981). This topic has been reviewed in the past and a choice of 3 for RBE was considered reasonable for uranium miner dosimetry (Parker, 1969; Richmond and Boecker, 1971; Walsh, 1971). A more recent paper based upon human lung cancer experience supports a value of about 5 (Walsh, 1979).

4.6 Summary

In conclusion, experimental data on uranium mine atmospheres and physiological parameters are necessary to derive an accurate exposure limit based upon alpha dose to the bronchial tree for radon daughters.

The least data exist for the biological clearance of the tracheobronchial region and the particle size distribution of the inhaled aerosol. Other variables, such as radon daughter disequilibrium, correction for non-laminar flow in the larger bronchi, weighting dose by number of cells at risk, etc., appear to introduce differences by factors of perhaps 2 (often compensatory to one another). This is discussed in Section 6. Calculated differences due to clearance are shown in Table 4.2 to affect only one airway generation of the tracheobronchial tree significantly, and that only if no clearance is assumed. Except for very unusual exposure conditions, the WLM standard may adequately describe the risk from uranium mine aerosols, but incorporation of a factor into the WL definition to account for activity that is unattached to carrier aerosols will improve the accuracy of the WL unit as a measure of risk (Cross, 1979). Finally, attention must be given to the methodology of dose averaging. Average tracheobronchial dose is not sufficient to define the risk of exposure to uranium mine aerosols. Regional doses (even as small as a single generation) are more consistent with present radiation protection philosophy.

5. Dosimetry of Inhaled Radon Daughters in Environmental Atmosphere

5.1 Introduction

Exposure to radon daughters in normal environmental situations differs from that in mining atmospheres in three respects: *1*) the median particle size of the carrier aerosol is smaller (0.1 μm versus about 0.2 to 0.4 μm for mines, George, 1975; George and Breslin, 1980); *2*) the fraction of ^{218}Po(RaA) that exits as free ions is larger (.07 versus .04 in mines, George, 1975, 1978); and *3*) the exposure is continuous rather than part time. All three of these factors increase the bronchial dose per unit exposure. Compensating for these effects, the environmental concentration of radon daughters is usually lower than in an underground situation.

The purpose of this section is to indicate current estimates for environmental dose conversion factors and to apply them to recently published values for environmental radon daughter concentrations. Also, since an environmental population is composed of men, women, and children, estimated bronchial dose conversion factors for all three groups are given. The external whole body gamma ray dose from radon daughters should also be mentioned here (9×10^{-6} rad/year per pCi/m^3, Beck, 1974), but since it is small relative to the alpha dose from inhalation, it will not be considered further.

5.2 Required Data for the Dose Formulation

Dose formulation requires a knowledge of the deposition, clearance and physical buildup and decay of the radon daughters. Some of these factors, in, turn depend on such items as lung morphometry and breathing patterns. The deposition of aerosols within the bronchial tree is reasonably well predicted by the Gormley-Kennedy (1949) diffusion equations once laminar or Poiseuille flow has been established in the airways. This is discussed in detail in Section 6. One serious objection to their use is that turbulence exists in about the first five airway generations, and this should produce increased particle deposition. The magnitude of this increased deposition, currently termed convective diffusion, has been studied by several investigators (Martin and Jacobi, 1972; Cheng, 1973; Chan, 1978). The most complete data developed so far are those of Martin and Jacobi (1972). They determined deposition in hollow models of the bronchial tree from 0.2 μm natural aerosols tagged with ^{212}Pb. The cast dimensions were chosen to conform with the Weibel (1963) dichotomous Model A. Turbulence was established during inspiratory flow by introducing air to the cast through a right angle bend to simulate the nose and larynx. They found that, in the first six airway generations, deposition can be calculated by multiplying the Gormley-Kennedy value by the factors shown in Table 5.1.

A steady-state alpha activity per unit area is established on the bronchial tree as a result of particle deposition, clearance or removal by the mucociliary escalator, and radioactive build-up and decay of the daughters. This steady-state activity is usually found to be highest in generation 4 by a factor of 2 or so, and this corresponds roughly to the region where the majority of bronchial tumors are found (Schlesinger and Lippmann, 1978).

The correction for convective diffusion increases the steady-state activity in generation 4 for the daughters attached to the ambient aerosol. The total deposition of particles on the entire bronchial tree is only a few percent. On the other hand, the unattached ^{218}Po(RaA) which has a very small particle size (~.001–0.010 μm) deposits completely on the bronchial tree. The correction for convective diffusion removes enough of the ions in the first three generations so that there

TABLE 5.1—*Correction factors developed by Martin and Jacobi (1972) to account for convective diffusion in upper airways. Gormley-Kennedy diffusion deposition, multiplied by correction factor yields true deposition. Correction factors apply for both inhalation and exhalation*

Generation	Location	Correction Factor
0	Trachea	7.5
1	Main Bronchi	5.5
2	Secondary Bronchi	5.0
3	Secondary Bronchi	4.0
4	Tertiary Bronchi	2.0
5	Tertiary Bronchi	1.5
All succeeding generations		1.0

is actually a smaller amount available for deposition in generation 4. The resulting steady-state or equilibrium activity for the unattached fraction is, therefore, somewhat smaller in generation 4 when convective diffusion is accounted for. The net effect of convective diffusion upon the total dose is small but merely shifts the apportionment of dose between the attached and unattached fractions. In this section, the Martin and Jacobi corrections are applied, since they produce a result that is in better accord with actual lung dynamics, even though the corrected value of alpha dose differs only by about 30%.

The data necessary to complete the calculation of steady-state activity are the airway dimensions, clearance times, and breathing patterns. The data utilized for lung morphometry are based on the Weibel (1963) dichotomous Model A. The female bronchial tree and the tree of a ten-year-old child are scaled for the Weibel morphometry using the following assumptions. The female tree is scaled assuming that both the cross-sectional airway area and the lengths are 0.9 of the male Weibel model (volume = 0.8 of male). Some support for this is found in the ratio of the female/male dead space which is 130 mL/160 mL = 0.8 (ICRP, 1975). The bronchial tree dimensions of the ten-year-old child are estimated from data given in ICRP Publication 23 (ICRP, 1974) which indicates that the diameter and length of the trachea and major bronchi are one-half those of the Weibel model. It is assumed that all airways in the ten-year-old can be scaled using this factor. No detailed morphometric data for women or children are known to exist.

The breathing patterns for males, females, and the ten-year-old are adopted from ICRP Publication 23. The breathing patterns and lung morphometry are listed in Tables 5.2 and 5.3.

TABLE 5.2—*Breathing parameters for the male, female and ten-year-old child*[a]

	Male	Female	Child (10 Years)
Light activity (16 hours per day)			
Breaths per minute	15	15	24
Liters per breath	1.25	0.94	0.60
Liters per minute	18.75	14.1	14.4
Liters in 16 hours	18000	13536	13824
Resting (8 hours per day)			
Breaths per minute	12	12	16
Liters per breath	0.75	0.40	0.30
Liters per minute	9	4.8	4.8
Liters in 8 hours	4320	2304	2304
Total volume (liters per day)	22320	15840	16128

[a] Breathing cycle for all calculations of velocity is based on Landahl (1950). Inspiration is 3/8 cycle time, expiration is 3/8 cycle time, pause after inspiration or expiration, 1/8 cycle time. Cycle time in seconds equals 60/breaths per minute.

TABLE 5.3—*Lung morphometry based on Weibel dichotomous Model A. Female airways scaled so that volume is 0.80 of male, child airways scaled to 1/2 of adult male dimensions. All dimensions in centimeters*

Airway Generation	Mucus Transit[a] Time	Male		Female		Child (10 Years)	
		Airway Radius	Airway Length	Airway Radius	Airway Length	Airway Radius	Airway Length
0	8 Minutes	0.900	12.0	0.855	10.8	0.450	6.00
1	6	0.610	4.76	0.580	4.28	0.305	2.38
2	8	0.415	1.90	0.390	1.71	0.2075	0.95
3	3	0.280	0.76	0.270	0.68	0.140	0.38
4	14	0.225	1.27	0.210	1.14	0.1125	0.635
5	12	0.175	1.07	0.170	0.96	0.0875	0.535
6	10	0.140	0.90	0.130	0.81	0.0700	0.45
7	32	0.115	0.76	0.110	0.68	0.0575	0.38
8	27	0.093	0.64	0.088	0.58	0.0465	0.32
9	23	0.077	0.54	0.073	0.48	0.0385	0.27
10	445	0.065	0.46	0.061	0.41	0.0325	0.23
11	378	0.0545	0.39	0.052	0.35	0.0272	0.195
12	320	0.0475	0.33	0.045	0.30	0.0237	0.165
13	261	0.0410	0.27	0.039	0.24	0.0205	0.135
14	223	0.037	0.23	0.035	0.21	0.0185	0.115
15	194	0.033	0.20	0.031	0.18	0.0165	0.100
16	160	0.030	0.165	0.028	0.148	0.0150	0.0825
17		0.027	0.141	0.026	0.127	0.0135	0.0705
18		0.0250	0.117	0.024	0.105	0.0125	0.0585
19		0.0235	0.099	0.022	0.089	0.0117	0.0495
20		0.0225	0.083	0.021	0.075	0.0112	0.0415
21		0.0215	0.070	0.020	0.063	0.0107	0.0350
22		0.0205	0.059	0.019	0.053	0.0102	0.0295
23		0.0205	0.050	0.019	0.045	0.0102	0.0250

[a] Based on values reported by Altshuler *et al.* (1964). Mucus transit times are considered to be the same for all adults and for the ten-year-old child 1/2 the value listed.

The steady state airway alpha activity in Weibel generation 4 calculated with the data of Tables 5.1 and 5.3 is shown in Table 5.4. The particle size of the ambient aerosol is taken as 0.125 μm activity median diameter, the value measured by George and Breslin (1980). The alpha equilibrium activity is calculated for men, women, and the ten-year-old child for both light activity and resting breathing patterns.

It is evident from Table 5.4 that the different breathing patterns of a given individual do not result in equilibrium activity concentrations that differ by even a factor of two for the attached fraction of the daughters. Martin and Jacobi (1972) showed that the Gormley-Kennedy equations may be approximated by the function;

$$\text{fractional airway deposition} = k\, Q^{-2/3} \qquad (5\text{-}1)$$

where

k = a constant depending upon the airway, and
Q = inspiratory flow rate in liters per minute in a given airway.

TABLE 5.4—*Steady state alpha activity (pCi/cm²) in Weibel Generation 4 for atmospheric concentration of 1000 pCi/m³ of each of the inhaled nuclides, unattached[a] RaA, attached[b] RaA, RaB and RaC*

Equilibrium Activity	Inhaled Nuclide							
	RaA*		RaA		RaB		RaC	
	RaA	RaC′	RaA	RaC′	RaA	RaC′	RaA	RaC′
Male								
Light activity	0.056	0.030	0.0013	0.0011	—	0.0096	—	0.0082
Resting	0.020	0.0086	0.0010	0.0008	—	0.0073	—	0.0063
Female								
Light Activity	0.049	0.024	0.0013	0.0011	—	0.0010	—	0.0083
Resting	0.006	0.002	0.0008	0.0007	—	0.0061	—	0.0053
Child (10 Years)								
Light Activity	0.16	0.058	0.0031	0.0018	—	0.016	—	0.017
Resting	0.043	0.010	0.0020	0.0011	—	0.010	—	0.011

[a] Unattached RaA (RaA*) is calculated assuming the diffusion coefficient to be 0.054 cm²/sec at ambient temperature or 0.055 cm²/sec corrected to body temperature.

[b] Carrier aerosol for attached fraction of radon daughters is 0.125 μm activity median diameter. Nose breathing assumed. Nasal deposition 60% for RaA* and 1.3% for attached daughters.

Even though more radon daughters are inhaled during physical activity because of a larger minute volume, their fractional deposition is reduced because of the higher flow rates. This compensation tends to keep the dust loading on the upper airways more uniform. Variations of equilibrium activity with breathing pattern for unattached RaA* is much larger since deposition of ions is very sensitive to flow rate.[2] At resting flow rates, the removal in the first three airway generations is substantial enough to affect the amount available for deposition in generation 4. This causes the steady-state activities for a resting breathing pattern to be effectively much smaller than that for light activity.

Steady-state activity for the child's brochial tree is substantially higher than that for men or women. This is due primarily to the reduced airway area. The total activity in the airways is roughly the same but the small airway area results in a larger equilibrium activity per unit area.

5.3 The Dose Calculation

The alpha dose is estimated for 22 μm below the surface of the bronchial epithelium. This is the location of the shallowest basal cell nuclei which are thought to be implicated in the etiology of bronchogenic carcinoma. Spatial dose calculations result in dose conversion factors at 22 μm of 8.8 rad/year per pCi/cm² for RaA and 16 rad/year

[2] In all Tables RaA* refers to unattached or ionic RaA.

per pCi/cm^2 for RaC′ on the bronchial epithelium. The dose conversion factors are discussed in detail in Section 6.

The dose conversion factors may be combined with the values of steady-state activity on the bronchial tree for each radon daughter with unit activity in the atmosphere to yield the following expressions for annual dose (mrad/y) for men, women and ten-year-old child. Note that RaA* is unattached RaA concentration.

Male

Light activity, dose = 0.98(pCi RaA*/m^3) + 0.029(pCi RaA/m^3) + 0.16(pCi RaB/m^3) + 0.14(pCi RaC/m^3) (5-2)

Resting, dose = 0.32(pCi RaA*/m^3) + 0.022(pCi RaA/m^3) + 0.12(pCi RaB/m^3) + 0.10(pCi RaC/m^3) (5-3)

Female

Light activity, dose = 0.82(pCi RaA*/m^3) + 0.029(pCi RaA/m^3) + 0.16(pCi RaB/m^3) + 0.14(pCi RaC/m^3) (5-4)

Resting, dose = 0.29(pCi RaA*/m^3) + 0.019(pCi RaA/m^3) + 0.10(pCi RaB/m^3) + 0.09(pCi RaC/m^3) (5-5)

Child (ten-year-old)

Light activity, dose = 2.36(pCi RaA*/m^3) + 0.06(pCi RaA/m^3) + 0.26(pCi RaB/m^3) + 0.28(pCi RaC/m^3) (5-6)

Resting, dose = 0.54(pCi RaA*/m^3) + 0.04(pCi RaA/m^3) + 0.17(pCi RaB/m^3) + 0.18(pCi RaC/m^3) (5-7)

Unfortunately, detailed characterizations of radon daughters in the environment are scarce. Therefore, only limited estimation of bronchial dose with Eqs. (5-2) to (5-7) is possible. Generally, the approximations used are that a fixed activity ratio of the daughters exists with respect to radon and that the fraction of unattached RaA is constant.

Several studies have shown (NCRP Report No. 45, NCRP, 1975) that for indoor and outdoor atmospheres the ratio of radon to its short-lived daughters is 1/0.5/0.3/0.2 and 1/0.9/0.7/0.7 (^{222}Rn/RaA/RaB/RaC) respectively. One study of residences in the New York, New Jersey area indicates that the fraction of unattached RaA, that is, RaA*/^{222}Rn equals 0.07 (George and Breslin, 1980). This is in agreement with a previous study (Fisenne and Harley, 1974) where

0.085 was measured as an average outdoor ratio of RaA*/^{222}Rn and 0.07 was measured as the indoor ratio.

Assuming a value of 0.07 for the fraction of unattached RaA and an equilibrium activity ratio of 1/0.9/0.7/0.7, Eqs. (5.2) to (5.7) may be expressed in terms of the ^{222}Rn concentration alone with the following expressions:

Male

Dose

Light activity(mrad/year) =0.30(pCi ^{222}Rn/m^3) (5-8)
Resting(mrad/year) =0.20(pCi ^{222}Rn/m^3) (5-9)

Female

Dose

Light activity(mrad/year) =0.29(pCi ^{222}Rn/m^3) (5-10)
Resting(mrad/year) =0.15(pCi ^{222}Rn/m^3) (5-11)

Child (ten-year-old)

Dose

Light activity(mrad/year) =0.59(pCi ^{222}Rn/m^3) (5-12)
Resting(mrad/year) =0.31(pCi ^{222}Rn/m^3) (5-13)

The required detail in estimating population exposure can never be known if only approximations of the detailed atmospheric characteristics are used in dose estimations. In particular, unattached RaA needs to be measured since it contributes substantially to the dose. Future studies should be designed to provide sufficient characterization of the atmospheric aerosol so that a range of values for population dose can be determined, rather than a broad average.

5.4 Population Dose Estimates

A few detailed studies of radon or radon daughters in the environment are available (Hultqvist, 1956; Toth, 1972; Steinhausler *et al.*, 1978; Cliff, 1978; Stranden *et al.*, 1979; George and Breslin, 1980; Swedjemark, 1980; McGregor *et al.*, 1980). Cliff (1978) has measured the total RaA concentration in outside air over five different classifications of subsoil in Great Britain. His values averaged 50, 70, 90, 75, and 50 pCi RaA/m^3 for chalk, clay, granite, limestone, and sandstone, respectively. He found that the radon flux inside 87 dwellings constructed of five different types of mainshell material was constant within a factor of about two. The ventilation rate of the dwelling was the major source of difference among structures. He also estimated the annual exposure in Great Britain as 0.15 WLM/year.

Steinhausler *et al.*, (1978) measured ^{222}Rn, ^{220}Rn; RaA, RaB, RaC, RaD, and ThB in five different buildings over a one year period to assess the dose to persons living in Salzburg, Austria. The average outdoor concentration at these sites was found to be 220 pCi ^{222}Rn/m^3 and the indoor average was 600 pCi ^{222}Rn/m^3. Their average value for radon daughter equilibrium indoors was 1/0.9/0.6/0.4.

McGregor *et al.* (1980) measured radon and radon daughters in Canadian homes during the summer in 19 cities to study the distribution of exposure. The median ^{222}Rn concentration and radon daughter exposure levels were 500 pCi/m^3 and 0.003 WL, respectively.

In the United States, George and Breslin (1980) measured the distribution of radon and radon daughter concentrations in 21 New York and New Jersey residences. Most of the dwellings were single-family two-story wood frame or brick construction. The radon concentrations measured over a period of several years were found to have a lognormal distribution for both indoor and outdoor concentrations. The geometric means for outdoor, first floor, indoor, and cellar concentrations were 180, 830, and 1700 pCi ^{222}Rn/m^3, respectively. The unattached fraction of RaA was found to be 0.07 (RaA*/^{222}Rn). They also measured the Working Levels (WL) in these same locations and these were also found to be distributed lognormally, with geometric means for outdoor, first floor, and cellar of 0.0016, 0.0041, and 0.0081 WL. These data are shown in Figs. 5.1 and 5.2. It is of interest that they found an average value of WL per pCi ^{222}Rn per cubic meter to be 6×10^{-6}.

Swedjemark (1980), found that indoor radon concentrations in Swedish dwellings built before 1940 averaged 780 pCi ^{222}Rn/m^3, those built between 1940 and 1971 averaged 1900 pCi ^{222}Rn/m^3 and those built between 1971 and 1973 averaged 3200 pCi ^{222}Rn/m^3.

It was discussed in Section 1 that the main source of indoor radon and daughters is the soil under and around the structure and that the indoor levels are dependent primarily upon ventilation. The studies so far have provided data to this effect and with the exception of special sources of radon, such as elevated levels in the household water supply (Gesell and Prichard, 1980; Castren, 1980; Hess *et al.*, 1980) indoor radon and radon daughter concentrations display reasonable uniformity when ventilation is accounted for.

This section is concerned primarily with evaluating radon daughter population exposures in the United States. It appears, however, that

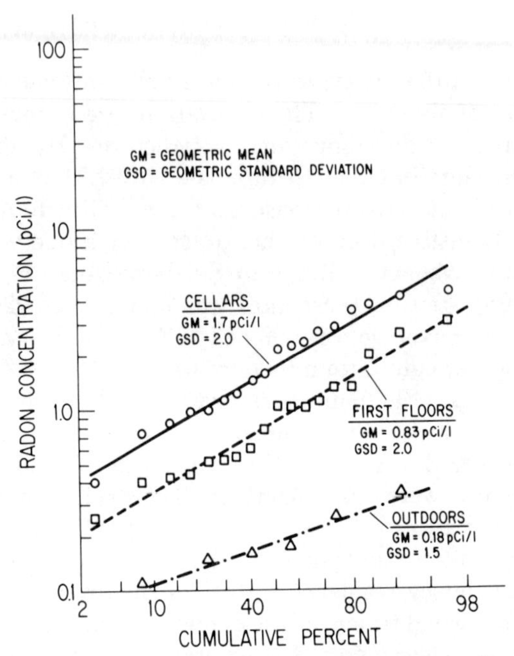

Fig. 5.1. Distribution of annual radon concentration (from George and Breslin, 1980).

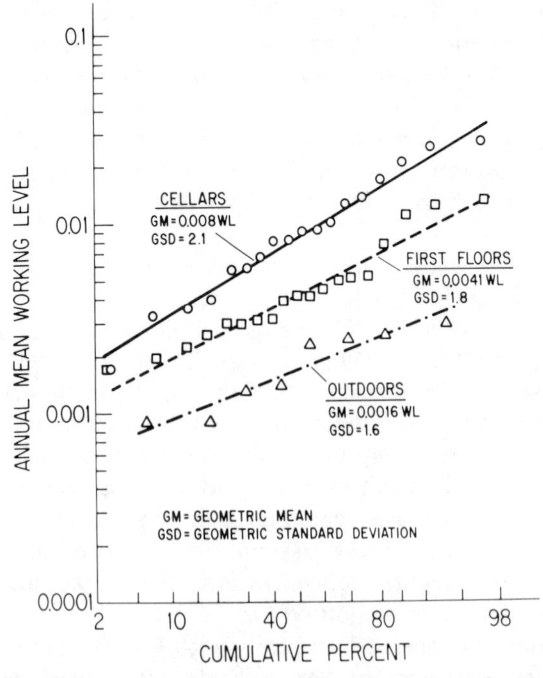

Fig. 5.2. Distribution of annual mean working level (from George and Breslin, 1980).

data from other countries are not substantially different, Sweden being a possible exception (Swedjemark, 1980), and data may probably be combined where available.

Approximate annual dose may be estimated if it is known how much time is spent either indoors or outdoors and the type of physical activity engaged in which determines the breathing pattern. For simplicity, it is assumed that the average person spends 16 hours a day engaged in light activity, 4 hours of which is outdoors, and 8 hours per day resting. It is assumed that a child ten years old spends 14 hours a day engaged in light activity, 6 hours of which is outdoors, and 10 hours per day resting. The annual dose formulae then may be reduced to:

Male

$$\text{Annual dose (mrads)} = 0.30(12/24)(\text{pCi } ^{222}\text{Rn}_\text{I}/\text{m}^3) + 0.20(8/24)(\text{pCi } ^{222}\text{Rn}_\text{I}/\text{m}^3) + 0.30(4/24)(\text{pCi } ^{222}\text{Rn}_0/\text{m}^3) = 0.22(\text{pCi } ^{222}\text{Rn}_\text{I}/\text{m}^3) + 0.05(\text{pCi } ^{222}\text{Rn}_0/\text{m}^3) \quad (5\text{-}14)$$

Female

$$\text{Annual dose (mrads)} = 0.29(12/24)(\text{pCi } ^{222}\text{Rn}_\text{I}/\text{m}^3) + 0.15(8/24)(\text{pCi } ^{222}\text{Rn}_\text{I}/\text{m}^3) + 0.29(4/24)(\text{pCi } ^{222}\text{Rn}_0/\text{m}^3) = 0.20(\text{pCi } ^{222}\text{Rn}_\text{I}/\text{m}^3) + 0.05(\text{pCi } ^{222}\text{Rn}_0/\text{m}^3) \quad (5\text{-}15)$$

Child (ten-year-old)

$$\text{Annual dose (mrads)} = 0.59(8/24)(\text{pCi } ^{222}\text{Rn}_\text{I}/\text{m}^3) + 0.31(10/24)(\text{pCi } ^{222}\text{Rn}_\text{I}/\text{m}^3) + 0.59(6/24)(\text{pCi } ^{222}\text{Rn}_0/\text{m}^3) = 0.32(\text{pCi } ^{222}\text{Rn}_\text{I}/\text{m}^3) + 0.15(\text{pCi } ^{222}\text{Rn}_0/\text{m}^3) \quad (5\text{-}16)$$

where $^{222}\text{Rn}_\text{I}$ and $^{222}\text{Rn}_0$ = indoor and outdoor radon concentration.

To estimate a population dose, we assume that the average indoor and outdoor concentrations of radon are the geometric mean values for the first floor and outdoors measured by George and Breslin (1980) of 830 pCi/m^3 and 180 pCi/m^3. This yields annual dose to basal cells in bronchial epithelium (generation 4 at 22 μ depth) of:

Male

$$\text{annual dose} = 190 \text{ mrad} \quad (5\text{-}17)$$

Female

$$\text{annual dose} = 170 \text{ mrad} \quad (5\text{-}18)$$

Child (ten-year-old)

$$\text{annual dose} = 300 \text{ mrad} \quad (5\text{-}19)$$

Since the annual dose to a ten year old child is almost a factor of two higher than for adults, it is of interest to estimate the dose to the infant to determine if this effect persists. The annual dose to a one-year-old infant was calculated by using the breathing parameters given in ICRP Publication 23 (ICRP, 1975) and by assuming the diameter and length of the airways to be one-third of the Weibel model. The annual dose to the one-year-old infant calculated in this way for the indicated radon concentrations and assuming 20 hours per day indoors and 4 hours per day outdoors, is 170 mrad or about the same as that for the adult female.

5.5. Environmental Alpha Dose Factors Using Yeh-Schum Bronchial Morphometry

As additional morphometric data become available it is of interest to utilize newer values for radon daughter dosimetry. In Section 6 it is explained that widely differing lung dimensions (child to adult) do not affect the calculated alpha dose by more than about 60% when Weibel morphometry is considered. Yeh and Schum (1980) have reported bronchial morphometry in a complete five-lobed adult male lung. Their values have been used to calculate the alpha dose for inhaled radon daughters under environmental conditions in units of rad/WLM in the airways of the adult male, adult female, ten-year-old child and one year old infant (Harley and Pasternack, 1982). The values are shown in Table 5.5 a, b, c, d. The highest alpha dose calculated using the Yeh-Schum morphometry occurs in generation 2 (major bronchi) and the average over the lobes for this airway is 0.7, 0.7, 1.7 and 1.3 rad/WLM in the adult male, adult female, ten-year-old child and one-year-old infant, respectively. If an average environmental exposure of 0.2 WLM/year is assumed, then the alpha dose in this airway for these persons is 140, 140, 340 and 260 mrad, which is in reasonable agreement with values obtained with Weibel morphometry in generation 4 of 190, 170, 300 and 170 mrad, respectively.

It should be pointed out that the alpha dose factor in rad/WLM for environmental exposure is somewhat higher than that for underground exposure. This is due to the higher values of unattached RaA measured environmentally and the lower environmental breathing rates with subsequent higher bronchial deposition of radon daughters.

TABLE 5.5a—*Alpha dose conversion factors (rad/WLM) for environmental radon daughters in the conducting airways in a five lobed human adult male lung. Atmospheric and respiratory characteristics given in the footnotes*[a]

Generations	Right Lobes			Left Lobes	
	Upper	Middle	Lower	Upper	Lower
0	0.37	—	—	—	—
1	0.38	—	—	0.34	—
2	0.80	0.36	—	0.87	0.89
3	0.30	0.15	0.28	0.26	0.22
4	0.41	0.45	0.54	0.63	0.59
5	0.20	0.18	0.24	0.31	0.39
6	0.14	0.15	0.11	0.16	0.17
7	0.22	0.24	0.29	0.37	0.36
8	0.15	0.17	0.18	0.17	0.21
9	0.09	0.10	0.06	0.11	0.08
10	0.33(0.46)	0.36(0.50)	0.16(0.22)	0.35(0.49)	0.22(0.30)
11	0.35(0.49)	0.46(0.64)	0.18(0.26)	0.39(0.56)	0.24(0.33)
12	0.40(0.57)	0.48(0.68)	0.22(0.31)	0.42(0.59)	0.29(0.40)
13	0.35(0.49)	0.51(0.71)	0.26(0.36)	0.36(0.51)	0.35(0.49)
14	—	0.45(0.63)	0.29(0.41)	—	0.32(0.45)
15	—	—	0.29(0.41)	—	—
16	—	—	0.22(0.30)	—	—

1. ^{222}Rn to radon daughter ratio: ^{222}Rn/RaA/RaB/RaC′ = 1/0.90/0.70/0.70.

2. Fraction of RaA that is not attached to aerosols is 0.07 times the ^{222}Rn concentration. Diffusion coefficient of unattached RaA equals 0.005 cm^2/sec.

3. The particle size (AMD) is 0.12 μm for the attached fraction of radon daughters.

4. Gormley-Kennedy diffusion deposition fractions are multiplied by factors determined by Martin and Jacobi to account for somewhat enhanced deposition in generations 1 to 6 from convective or turbulent diffusion during both inspiration and expiration.

5. Breathing rate for the active male is 18.75 L/min, 1250 cm^3 per breath, 4 second breathing cycle (3/8, 1/8, 3/8, 1/8) inspiration, pause, expiration, pause. Resting rate 9 L/min, 750 cm^3 per breath, 5 second cycle.

6. Mucus clearance times taken from.

7. Twenty percent alveolar deposition subtracted before calculation of deposition during expiration.

8. The alpha dose is that to shallow basal cells 22 μm below the surface of the bronchial epithelium. From generation 10 to 16 where epithelial thickness is thin the values in parentheses are the alpha dose at 10 μm below the surface of the bronchial epithelium.

9. Active and resting dose conversion factors combined to yield 16 hours per day active and 8 hours per day resting.

[a] From Harley and Pasternack, 1982.

TABLE 5.5b—*Alpha dose conversion factors (rad/WLM) for environmental radon daughters in the conducting airways in a five lobed human adult female lung. Respiratory and atmospheric characteristics given in the footnotes*[a]

Generations	Right Lobes			Left Lobes	
	Upper	Middle	Lower	Upper	Lower
0	0.36	—	—	—	—
1	0.36	—	—	0.32	—
2	0.76	0.34	—	0.83	0.84
3	0.28	0.14	0.26	0.25	0.21
4	0.39	0.42	0.51	0.60	0.56
5	0.19	0.18	0.23	0.29	0.37
6	0.14	0.14	0.11	0.15	0.16
7	0.21	0.22	0.28	0.35	0.34
8	0.14	0.16	0.17	0.16	0.20
9	0.08	0.10	0.06	0.10	0.08
10	0.31(0.43)	0.34(0.48)	0.15(0.21)	0.33(0.46)	0.20(0.29)
11	0.33(0.47)	0.43(0.61)	0.17(0.24)	0.37(0.52)	0.22(0.32)
12	0.38(0.54)	0.45(0.63)	0.21(0.29)	0.39(0.55)	0.27(0.38)
13	0.33(0.46)	0.48(0.67)	0.24(0.34)	0.34(0.48)	0.33(0.46)
14	—	0.42(0.60)	0.28(0.39)	—	0.30(0.42)
15	—	—	0.28(0.38)	—	—
16	—	—	0.21(0.29)	—	—

1. All parameters similar to those of Table 5.5a except breathing rate of the active female is 14.1 L/min, 940 cm^3 per breath with a 4 second breathing cycle. Breathing rate of the resting female is 4.8 L/min, 400 cm^3 per breath with a 5 second breathing cycle. The individual dose factors are combined to yield 16 hours per day active and 8 hours per day resting.

2. The conducting airways in the female lung are scaled from the adult male lung so that the volume is 0.80 of the male lung. Diameter of each airway is 0.94 of the male value and the length 0.90.

[a] From Harley and Pasternack, 1982.

TABLE 5.5c—*Alpha dose conversion factors (rad/WLM) for environmental radon daughters in the conducting airways in a five lobed human child's lung (10 y old). Respiratory and atmospheric characteristics given in the footnotes*[a]

Generations	Right Lobes			Left Lobes	
	Upper	Middle	Lower	Upper	Lower
0	0.85	—	—	—	—
1	0.85	—	—	0.76	—
2	1.81	0.81	—	1.99	2.02
3	0.67	0.34	0.63	0.59	0.50
4	0.94	1.01	1.22	1.43	1.34
5	0.46	0.42	0.54	0.70	0.89
6	0.33	0.33	0.26	0.37	0.39
7	0.50	0.54	0.66	0.84	0.82
8	0.34	0.40	0.40	0.39	0.47
9	0.21	0.24	0.15	0.25	0.19
10	0.75(1.06)	0.82(1.16)	0.36(0.51)	0.80(1.13)	0.49(0.70)
11	0.81(1.14)	1.06(1.49)	0.42(0.59)	0.90(1.28)	0.54(0.77)
12	0.94(1.32)	1.11(1.56)	0.51(0.71)	0.96(1.36)	0.66(0.93)
13	0.79(1.12)	1.17(1.64)	0.59(0.83)	0.84(1.19)	0.80(1.14)
14	—	1.02(1.44)	0.68(0.94)	—	0.72(1.02)
15	—	—	0.67(0.93)	—	—
16	—	—	0.50(0.70)	—	—

1. All parameters similar to those of Table 5.5a except breathing rate of the active child is 14.4 L/m, 600 cm^3 per breath, with a 2.5 second breathing cycle. Breathing rate of the resting child is 4.8 L/m, 300 cm^3 per breath, with a 3.75 second breathing cycle. The individual dose factors are combined to yield 16 hours per day active and 8 hours per day resting.

2. The ten-year-old child's lung is scaled from the adult male lung. The diameter and length of each airway is 0.5 that of the adult male.

[a] From Harley and Pasternack, 1982.

TABLE 5.5d—*Alpha dose conversion factors (rad/WLM) for environmental radon daughters in the conducting airways in a five lobed human infant lung (1 y old). Respiratory and atmospheric characteristics given in the footnotes*[a]

Generations	Right Lobes			Left Lobes	
	Upper	Middle	Lower	Upper	Lower
0	0.71	—	—	—	—
1	0.70	—	—	0.63	—
2	1.49	0.66	—	1.62	1.64
3	0.55	0.27	0.51	0.48	0.41
4	0.76	0.82	0.98	1.16	1.09
5	0.37	0.34	0.44	0.57	0.72
6	0.26	0.27	0.21	0.29	0.32
7	0.40	0.43	0.53	0.68	0.66
8	0.27	0.32	0.32	0.31	0.38
9	0.16	0.18	0.12	0.20	0.15
10	0.59(0.83)	0.65(0.90)	0.28(0.40)	0.63(0.88)	0.39(0.54)
11	0.64(0.89)	0.83(1.16)	0.33(0.46)	0.71(0.99)	0.43(0.60)
12	0.74(1.03)	0.88(1.22)	0.40(0.55)	0.75(1.05)	0.52(0.73)
13	0.63(0.88)	0.92(1.28)	0.47(0.65)	0.65(0.91)	0.63(0.88)
14	—	0.82(1.14)	0.54(0.75)	—	0.57(0.80)
15	—	—	0.54(0.74)	—	—
16	—	—	0.40(0.56)	—	—

1. All parameters similar to those of Table 5.5a, except the breathing rate of the infant is constant at 1.44 L/min, 48 cm^3 per breath, with a 2.0 second breathing cycle.
2. The infant lung is scaled from the adult male lung. The diameter and length of each airway is 0.333 that of the adult male.

[a] From Harley and Pasternack, 1982.

5.6 Summary

It is not known whether the levels used here to estimate population dose in the United States are really representative of the country as a whole. If they are, then the average dose over 50 years of adult life is about 10 rads for men and 9 rads for women. The additional dose during childhood would add a few rads to these values.

If the lifetime dose is distributed lognormally in the population, similar to the concentration values measured by George and Breslin (1980), then a few percent of the population can receive a lifetime bronchial dose three times these values through the normal variability in radon daughters.

This type of lifetime dose estimate should eventually be developed in some detail through measurement of the radon daughter characteristics in many geographic areas. With this information it is possible to estimate lung cancer risk for any population exposed to radon daughters and to determine which situations, if any, require remedial action.

6. Variability of Radon Daughter Dose Conversion Factors with Model Parameters

6.1 Introduction

Criticism of the calculation of radon daughter dose to cells in bronchial epithelium centers upon the fact that many of the biological and physical parameters must be estimated. It is felt by some that a dose model including ten variables cannot lead to a useful solution to the dose question. The ten variables are:

Physical Characteristics
1. Fraction of unattached RaA in the atmosphere
2. Daughter product equilibrium
3. Particle deposition models
4. Particle size distribution in the atmosphere
5. Physical dose calculation

Biological Characteristics
6. Breathing pattern (including nasal deposition)
7. Bronchial morphometry
8. Mucociliary clearance rate
9. Location of target cells
10. Mucus thickness

Two epidemiological findings justify the effort spent to determine the dose to bronchial epithelium. One is that the various studies which evaluate risk of bronchial cancer per WLM show consistency. Different types of underground mines located in widely different geographic regions are involved as well as different exposure levels of radon daughters.

The rough correspondence of cancer risk to air concentrations indicates that the bronchial dose relates to the radon daughter levels. The variability observed in the epidemiological data could be due to real variability in bronchial dose per WLM received by individual miners as well as to uncertainties in the exposure levels.

The second epidemiological finding is that bronchial tumors appear primarily in the upper airways at about the level of the third or fourth generation (lobar or segmental bronchi). Unless this region is special with regard to tumor induction, this indicates that the highest doses are delivered there. Since calculated doses are higher in this region by about a factor of two over other airway generations, some credibility is established in the dose modeling.

In this section, the variability of each of the factors indicated is discussed and the overall uncertainty in the dose estimate from each factor is calculated.

6.2 Fraction of Unattached RaA in the Atmosphere

Because of its efficient deposition in the upper bronchial tree, the dose from unattached RaA[3] can be 3 to 40 times that of each of the attached daughters per unit concentration in the atmosphere [see Eqs. (5-2) to (5-7)]. Until recently, the measurement of the unattached fraction was difficult and few data were available. Fortunately, in recent years several reliable measurement techniques have become available (James *et al.*, 1972; Thomas and Hinchliffe, 1972; George, 1972). Also, several published works indicate that it is possible to predict the value of the unattached fraction of RaA from a knowledge of the natural atmospheric aerosol concentration (condensation nuclei concentration) and the particle size.

Lassen and Rau (1960), Raabe (1969), Mohnen (1967), Duggan and Howell (1969a, 1969b) Kruger and Andrews (1976), Porstendorfer and Mercer (1978) and Ho *et al.* (1982) have described methods for calculating the degree of attachment of radon daughters to aerosols. One of the most extensive sets of measurements of unattached fraction as a function of aerosol particle size and concentration is reported by George *et al.* (1975) for uranium mine atmospheres. Their measurements along with the predictive equations of Raabe and Mohnen are shown in Figure 6.1 (Raabe, 1977). The difference between the measurements and the models is due primarily to the fact that the measured values result from the total effects of a particle size spectrum and the models each use a single particle size (Raabe, 1977). From Fig. 6.1 it

[3] Unattached RaA (RaA*) is defined in accordance with ICRP Publication 2 (ICRP, 1959) as the fraction of the equilibrium amount of RaA which is unattached to nuclei. (RaA*/^{222}Rn).

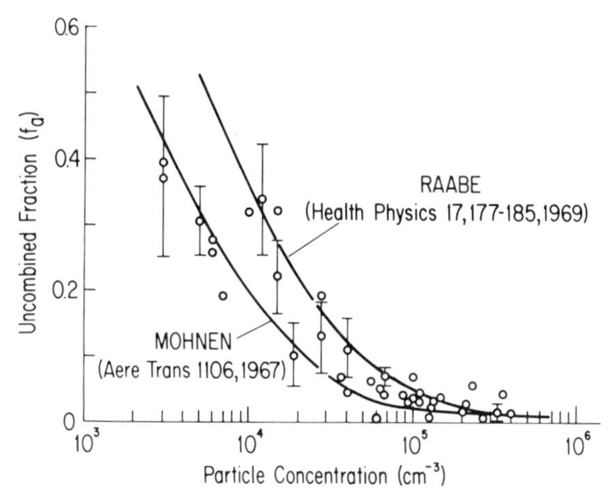

Fig. 6.1. Variation of uncombined (unattached) fraction of RaA with aerosol concentration as determined by George *et al.* (1975) and from the models of Mohnen (1967) and Rabe (1969).

can be seen that, in general, the predictive models agree well with the measurements. Since the unattached fraction is a function of particle concentration with secondary dependence on particle size, it is possible to estimate unattached RaA if a measurement of particle concentration is available.

One important conclusion that is evident from Fig. 6.1 is that there is a practical upper limit on the fraction of unattached RaA in most atmospheres since there is a minimum particle concentration which is generally observed. Rarely does the particle concentration in areas with any human activity fall below 10^3 to 10^4 particles per cm^3. One cigarette will profoundly affect the particle concentration indoors, and any human activity will increase the particle concentration several fold over the normal quiescent value. As an example of the typical particle concentration cycle, continuous measurements of condensation nuclei in a New York City basement laboratory and in a single family residence in Bayside (nonsmokers), New York are shown in Figs. 6.2 and 6.3 (George, 1978, unpublished data). In the early morning hours, particle concentrations subside but increase three to four fold with any human activity.

A value of 2×10^4 particles per cm^3 establishes an upper limit of 0.2 as a reasonable value of the unattached fraction of RaA. This maximum applies not only to environmental situations, but to uranium mines before 1960, prior to the use of diesel-powered equipment underground. The diesel exhaust fumes in recent uranium mining operations has resulted in very high particle concentrations. Measure-

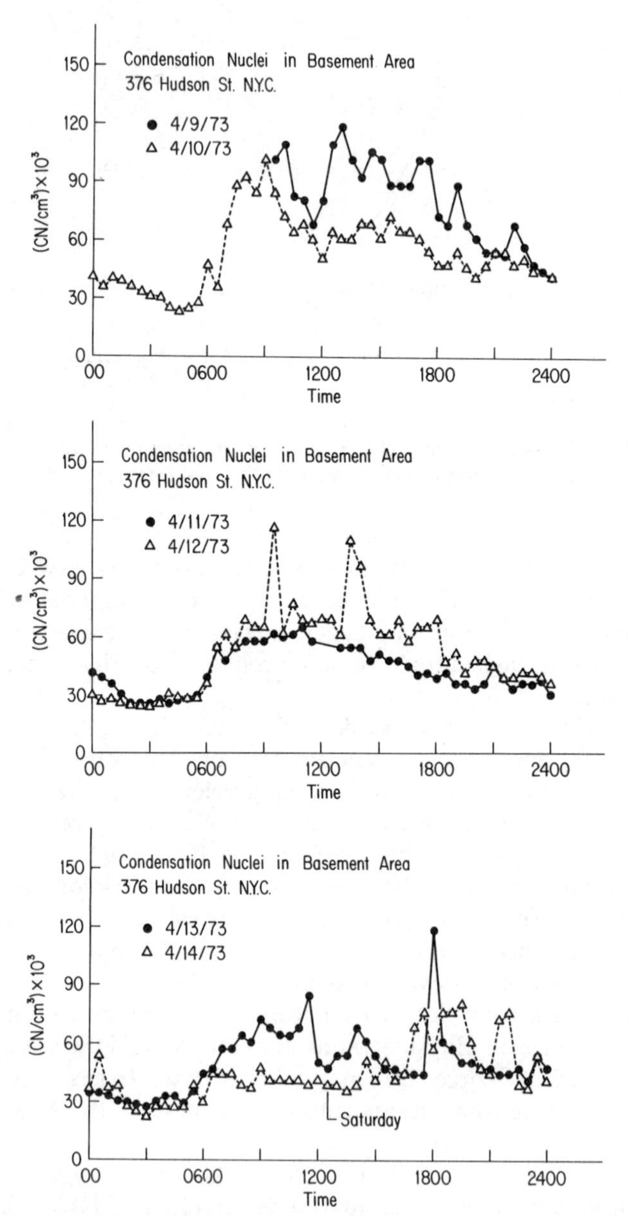

Fig. 6.2. Continuous measurements of condensation nuclei over a 12 day interval in New York City.

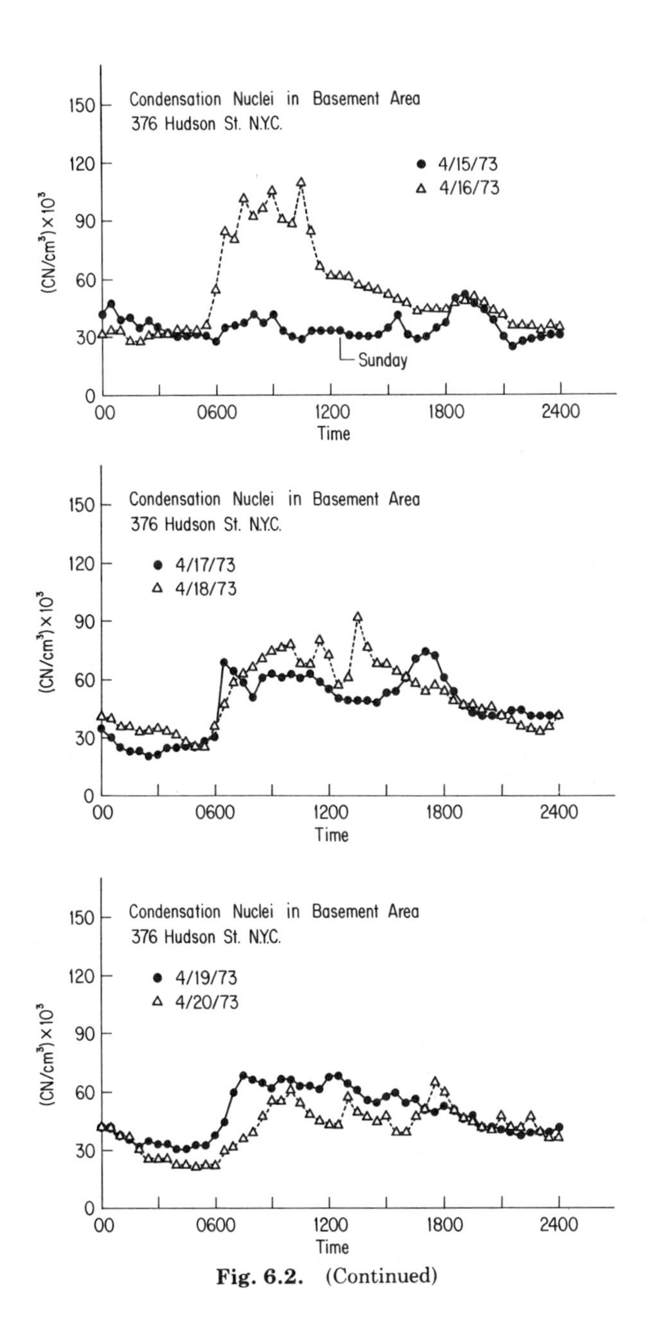

Fig. 6.2. (Continued)

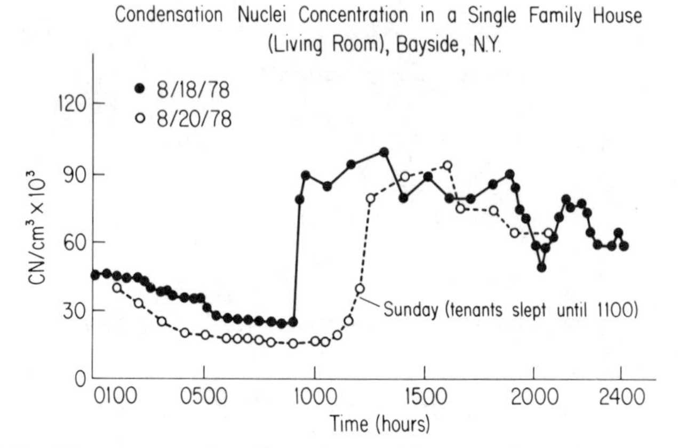

Fig. 6.3. Measurement of condensation nuclei in a single family dwelling in New York.

ments in 27 areas of four uranium mines with diesel-powered equipment showed an average unattached fraction of 0.04 (George *et al.*, 1975). The range of values was from 0.0004 to 0.16.

Although diesel exhaust fumes was not present in the older mines, it is difficult to believe that in actively mined areas the particle concentrations would be below 2×10^4 per cm^3.

The question also arises concerning possible levels of unattached RaB in the atmosphere. The main reason that dose conversion factors determined by Haque and Collinson (1967) are considerably higher than other estimates is their choice of free ion fractions of 0.3506, 0.0578, and 0.0771 for RaA, RaB, and RaC, respectively, in normal atmospheres.

Raabe (1969) has derived expressions to calculate free ion fractions for each of the daughters. These expressions apply to atmospheres that are in equilibrium (greater than 100 minutes old) and not to areas where ingrowth of the daughter products is occurring due to a fresh input of radon. The unattached fractions of RaA(f_a), RaB(f_b) and RaC(f_c) are given by:

$$f_a = \frac{\lambda_a}{\lambda_a + \lambda_s} \tag{6-1}$$

$$f_b = \frac{\lambda_b}{\lambda_b + \lambda_s}\,[f_a(1 - \alpha_a) + \alpha_a] \tag{6-2}$$

$$f_c = \frac{\lambda_c}{\lambda_c + \lambda_s}\,[f_b(1 - \alpha_b) + \alpha_b] \tag{6-3}$$

$$\lambda_s = \frac{S\bar{v}}{4} = \text{attachment coefficient} \tag{6-4}$$

$$S = n\pi\bar{D}_s^2 = \text{surface area concentration (cm}^2\text{/cm}^3\text{)} \tag{6-5}$$

where:

$\lambda_{a,b,c}$ = decay constants for RaA, RaB, RaC,
$\bar{D}_s$ = the average surface diameter of the aerosol particles for a given distribution of particles,
n = number concentration of aerosol particles per unit volume, of air
$\bar{v}$ = average velocity of the unattached daughters ($\bar{v} = 1.38 \times 10^4$ cm/sec), and
α_a = recoil fraction of attached RaA, which becomes unattached upon decay.

Mercer (1977) has calculated the fraction of attached RaA which becomes unattached upon decay (α_a) as 0.81 ± .07. This value has been used in Eqs. (6-1) to (6-3) in estimating the unattached fraction of radon daughters. Calculated values are shown in Table 6.1 for an average surface diameter $\bar{D}_s$ equal to 0.125 μm, the median value observed by George and Breslin (1980) in environmental atmospheres and for various particle concentrations. Mercer (1977) indicated that this calculated value for α_a of 0.81 could be less if the radioactive daughter atom has diffused into the carrier aerosol particle. For this reason, the values in Table 6.1 probably represent maximum values for the unattached fractions of RaB and RaC. George (1978) attempted to measure the unattached fraction of RaB in a uranium mine where there was an elevated concentration of the daughters. His measure-

TABLE 6.1—*Calculated values of unattached activity fractions of RaA (f_a), RaB (f_b) and RaC (f_c) for a particle size distribution with diameter of average surface equal to 0.125 μm. Atmosphere should be in equilibrium, i.e., greater than 100 minutes old.*

n(Particles cm^{-3})	$\lambda_s(min^{-1})$	f_a[a]	f_b[a]	f_c[a]
10^3	0.102	0.69	0.19	0.049
10^4	1.02	0.18	0.021	0.0007
2×10^4	2.03	0.10	0.010	0.0002
3×10^4	3.02	0.07	0.007	—
5×10^4	5.08	0.043	0.004	—
10^5	10.2	0.022	0.002	—
2×10^5	20.3	0.011	0.001	—
5×10^5	50.8	0.004	0.0004	—
10^6	102	0.002	0.0002	—

[a] $f_{a,b,c} = \dfrac{\text{Concentration of unattached RaA, RaB or RaC (pCi/m}^3)}{^{222}\text{Rn concentration (pCi/m}^3)}$

ments were never reported since the values obtained were indistinguishable from zero.

Measurements of unattached RaA in environmental situations indicate a mean value of 0.07. This would correspond to a particle concentration of 3×10^4 particles per cm^3 for the particle size used in Table 6.1. For these conditions, the calculated value for f_b is 0.007.

The calculated steady-state activity on the bronchial tree (pCi/cm^2) in Weibel Generation 4 for RaA*, RaB*, and RaC* (* indicates unattached form) in men, women, and for a ten-year-old child is given in Table 6.2. These steady-state alpha activities were calculated in the same manner as described in Section 5. The dose from RaA*, RaB*, and RaC* may be determined assuming that at 22 μm below the surface of bronchial epithelium the dose conversion factors are 8.8 rad/year per pCi/cm^2 for RaA and 16 rad/year per pCi/cm^2 for RaC′ in the mucous layer on the bronchial epithelium.

The annual alpha dose at 22 μm depth due to deposition of unattached RaB may then be estimated:

$$\text{Partial dose-RaB*(mrad/year)} = (A_{eq})(16)(f_b)(^{222}\text{Rn pCi/m}^3) \quad (6\text{-}6)$$

where

A_{eq} = steady state areal alpha activity (pCi/cm^2) in Weibel generation 4 as predicted from Table 6.2, and

f_b = fraction of ^{214}Pb (RaB) activity in unattached form (RaB*/^{222}Rn).

TABLE 6.2—*Steady state areal alpha activity (pCi/cm^2) in Weibel Generation 4 for atmospheric concentrations of 1000 pCi/m^3 of each of the inhaled unattached nuclides RaA*, RaB*, and RaC**[a]

Inhaled Nuclide	RaA*		RaB*		RaC*	
Equilibrium Activity	RaA	RaC′	RaA	RaC′	RaA	RaC′
Male						
Light Activity	0.056	0.030	—	0.28	—	0.30
Resting	0.020	0.008	—	0.079	—	0.095
Female						
Light Activity	0.048	0.024	—	0.23	—	0.25
Resting	0.006	0.002	—	0.018	—	0.025
Child (10 Years)						
Light Activity	0.16	0.058	—	0.55	—	0.74
Resting	0.043	0.010	—	0.10	—	0.17

[a] The deposition of the unattached nuclides RaA*, RaB* and RaC* is calculated using the Gormley-Kennedy diffusion equation corrected for convective diffusion with factors developed by Martin and Jacobi (1972). as in Section 5. The diffusion coefficient is assumed to be 0.054 cm^2/sec at ambient temperature and is corrected for body temperature of 37°C to 0.055 cm^2/sec. Nasal deposition is 60% for each unattached daughter.

If this dose is added to that of Eqs. (5-8) to (5-13), the total annual dose for a value of f_b of 0.007 would increase by 10 percent.

Although the effect of unattached RaB upon annual alpha dose for this case does not seem to be large, it remains an unresolved issue until measured values are available.

The range of unattached RaA values that are normally observed with either very high or low particle concentrations is from 0.04 – 0.2. This would cause a variation in the average annual dose (for males) calculated in Section 5 of −10 percent for 4 percent unattached RaA to +30 percent for 20 percent unattached RaA, respectively.

6.3 Daughter Product Disequilibrium

Radon daughter disequilibrium depends upon the age of the air mass and the interaction of the carrier aerosol with the materials and boundary surfaces present. When a submicron aerosol particle touches any surface, it is efficiently collected. Outdoor radon daughter ratios are generally higher than those observed indoors since the surface to volume ratio is smaller. Air conditioning filters or local air cleaning effectively remove radon daughters causing lower equilibrium ratios.

For a given radon concentration, a lower daughter ratio will give a smaller bronchial dose since fewer daughters are present. Examination of Eqs. (5-2) to (5-7) shows that if a typical environmental daughter activity ratio of ^{222}Rn/RaA/RaB/RaC of, say, 1/0.9/0.7/0.7 is compared with that of severe disequilibrium, say, 1/0.6/0.3/0.2, the reduction in dose for unit radon concentration is about 50 percent. Less severe disequilibrium, such as that reported by Steinhausler *et al.* (1978) for indoor atmospheres (1/0.9/0.6/0.4) produces a reduction in dose of 20 percent. Thus, the range of variation introduced into the dose calculation by radon daughter disequilibrium is about 20 percent for normal environmental situations.

6.4 Particle Deposition Models

Three different equations have been proposed as solutions to the diffusion equation for Poiseuille flow giving submicron aerosol loss from streams flowing through cylindrical tubes. Davies (1946) was the first to solve the diffusion equation and obtained:

$$C/C_0 = 0.819 \exp(-14.63\Delta) + 0.0926 \exp(-89.22\Delta) + 0.019 \exp(-212\Delta) + \cdots \tag{6-7}$$

where,

C = mean concentration of aerosol in the stream,
C_0 = initial concentration of aerosol in the stream,
Δ = diffusion parameter = $DL/4(Ur^2)$ (dimensionless),
D = diffusion coefficient (cm^2/sec),
L = length of tube (cm),
U = mean axial velocity of fluid (cm/sec), and
r = tube radius (cm).

Gormley and Kennedy (1949) solved the diffusion equation and obtained

$$C/C_0 = 0.819 \exp(-14.63\Delta) + 0.0976 \exp(-89.22\Delta) + 0.0325 \exp(-228\Delta) + \cdots \quad (6\text{-}8)$$

If Δ is small (<0.0078), then the solution given by 6-7 or 6-8 requires more terms in the series. Gormley and Kennedy gave an asymptotic solution to avoid this difficulty. Their solution for $\Delta < 0.0078$ is:

$$C/C_0 = 1 - 6.14\Delta^{2/3} + 4.8\Delta + 1.123\Delta^{4/3} + \cdots \quad (6\text{-}9)$$

Ingham (1975) has recomputed both the Davies and the Gormley-Kennedy solutions and believes that the expression given by Gormley-Kennedy is correct while the third term of Davies' solution contains a slight error. He derived an empirical expression which is identical with the Gormley-Kennedy solution but one which avoids the problem of computation with two equations.

Ingham (1975) suggests the use of the formula:

$$C/C_0 = 0.819 \exp(-14.63\Delta) + 0.0976 \exp(-89.22) + 0.0325 \exp(-228\Delta) + a \exp(-b\Delta^{2/3}) \quad (6\text{-}10)$$

where a and b are constants,

$$a = 0.0509, \text{ and}$$
$$b = 125.9.$$

Although there is no extensive experimental validation of any of the equations for diffusion deposition in bronchial airways, some supportive evidence exists. Martin and Jacobi (1972) found that, in a model of the bronchial tree based on the morphometry of Weibel, the Gormley-Kennedy equations predicted deposition once the flow became Poiseuille. In the more proximal airways, deposition could be calculated by multiplying the Gormley-Kennedy equations by a correction factor which depends on the airway (see Table 5.1). James (1976) measured the bronchial deposition of ^{212}Pb(ThB) in both unattached and attached form in ventilated pigs' lungs. When the lungs were ventilated without an artificial larynx at the entrance to the trachea, the deposition agreed well with the calculated values using the expression derived by Davies (1973). When an artificial "human" larynx was added to effect turbulence, the deposition of the attached aerosol (0.14 μm diameter) increased somewhat in the upper airways but was 1/3 to 1/2 of that observed by Martin and Jacobi (1972) for the human model. At about generation 4 of the pig lung, deposition could be predicted by the Davies equation. For the deposition of the unattached aerosol, James also found enhanced deposition in the first few airway generations but again less than that found by Martin and Jacobi. Also, the magnitude of the deposition in generations 4 through 7 was about 1/4 of that predicted by Davies. He indicated that this was probably due to rapid hygroscopic growth of particles in the bronchial tree to a size with an equivalent diffusion coefficient of 0.01 cm^2/sec. The commonly used value for the diffusion coefficient of unattached radon daughters is 0.054 cm^2/sec as determined by Chamberlain and Dyson (1956). The calculations in this present study, except where noted, use the value of 0.054 cm^2/sec corrected to body temperature of 37°C giving a value of 0.055 cm^2/sec.

The early work of Chamberlain and Dyson (1956) also gave experimental evidence that the Gormley-Kennedy equations were valid in straight tubes. Although their case of a human bronchial tree provided deposition data only for the trachea, it did indicate enhanced deposition over an ideal tube model.

Chan (1978) measured deposition of Fe_2O_3 particles (mass median aerodynamic diameter 3.5, with geometric standard derivation of 1.8) tagged with ^{99m}Tc in a hollow cast of the human bronchial tree. The cast included a larynx and extended to the sixth generation. Deposition in the first three generations agreed well with the values predicted by the Gormley-Kennedy equations corrected by the Martin-Jacobi factors. Deposition in the more distal airways was higher than that predicted by diffusion but the large value of σ_g makes it difficult to interpret his data. Since a considerable fraction of the radioactivity was associated with larger particle sizes (in the region where impaction would be important), the smaller diffusion deposition may have been masked.

From these studies it is evident that the Gormley-Kennedy equations appear adequate but that enhanced deposition does occur in the first few airways. The work of James (1976) with pig lung indicates the correction may not be more than a factor of two. However, the pig bronchial tree has markedly different morphometry compared with the human. A further complication is that the pig trachea has a small bronchus leading to an apical lobe above the main bronchial bifurcation. Both factors may reduce turbulence. On the other hand, the work of Chan (1978), using casts of the human bronchial tree, indicated that Martin and Jacobi (1972) correction factors may be directly applicable to humans.

These somewhat disparate experimental results do not produce the dilemma that they might. The reason for this is that more than half of the equilibrium activity at the 4th generation arises from activity deposited in generation 6 and beyond where the flow is laminar. The exception to this is the equilibrium activity of RaA itself. Because of its short-life, 70 percent of the RaA activity arises from material deposited in generation 4. The transit time in the upper airways is short due to mucociliary clearance and so the overall effect for the other daughters and RaC′ arising from RaA is minimized.

For the attached fraction of radon daughters, the equilibrium activity in generation 4 is increased by about 30 percent due to enhanced deposition. For the unattached fraction of RaA, the enhanced deposition in the trachea and major bronchi predicted by Martin and Jacobi is large. The equilibrium activity in generation 4 is thus reduced and, in fact, is about the same as if a larger particle size ($\sim D = 0.01$ cm^2/sec) were present.

6.5 Particle Size Spectrum

The particle size distribution of radon daughter activity in environmental or occupational situations has not been extensively studied. One of the most comprehensive environmental measurements was performed by George and Breslin (1980) in three New York and New Jersey residences. Their data are shown in Fig. 6.4. These measurements were performed with high flow rate diffusion batteries (Sinclair and Hoopes, 1975). The activity median particle diameter was found to be 0.125 μm. The distributions shown in Fig. 6.4 are typically bimodal and the small-sized mode is the unattached RaA.

George *et al.* (1975) also measured the size distribution of radon daughter bearing aerosols in four uranium mines near Grants, New Mexico. They found the activity median diameter at 27 locations ranged from 0.09 to 0.3 μm with a mean of 0.17 μm. The size distribution for the radon daughter aerosols was found to be lognormal. The geometric standard deviations observed from measurements ranged from 1.3 to 4 with a mean of 2.7.

The equilibrium alpha activity is calculated for several values of particle size in Weibel Generation 4 in Table 6.3. The particle sizes chosen are for unattached RaA, where the effective diameter is about 0.001 μm and for 0.05, 0.125, 0.17 μm, and 0.3 μm diameter aerosols. Excluding the effect of the unattached RaA, the equilibrium alpha activity varies moderately over this range of particle size. The use of a single particle size as representative of the atmospheric aerosol, if the aerosol has a lognormal distribution with large σ_g, probably introduces the most significant error in the dose calculation. The percent variation in annual dose for a very large median particle size shift from an average of 0.125 μm would range from +96 percent for the 0.05 μm particle to −20 percent for the 0.17 μm particle. Usually,

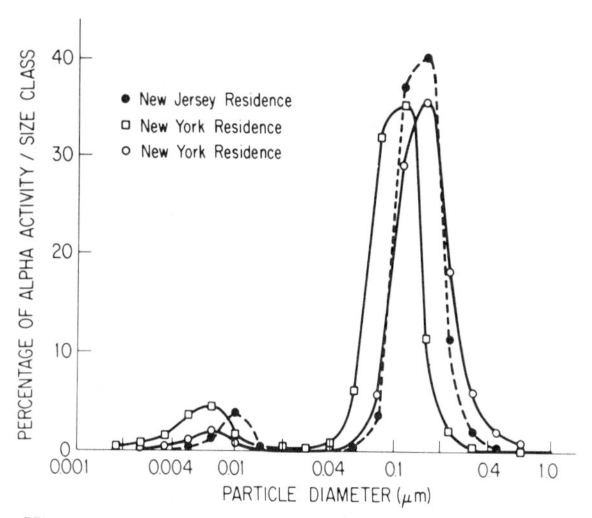

Fig. 6.4. Histograms of radon daughter size distribution (from George and Breslin 1980).

TABLE 6.3—*Steady state areal alpha activity in Weibel Generation 4 (pCi/cm²) for various particle sizes of inhaled carrier aerosol. Atmospheric concentration of each of the inhaled nuclides, unattached RaA, attached RaA, RaB and RaC is 1000 pCi/m³. Breathing pattern is that of the active male.*

Inhaled Nuclide	RaA*		RaA		RaB		RaC	
Equilibrium Activity	RaA	RaC′	RaA	RaC′	RaA	RaC′	RaA	RaC′
Particle Size of Carrier Aerosol								
~ 0.001 μm (Unattached RaA)	0.056	0.030						
0.05 μm			0.0031	0.0025	—	0.023	—	0.019
0.125 μm			0.0013	0.0011	—	0.0096	—	0.0087
0.17 μm			0.00097	0.00081	—	0.0072	—	0.0062
0.30 μm			0.00063	0.00052	—	0.0046	—	0.0040

however, some indication of particle size is available. The calculated alpha dose to other generations in the tracheobronchial tree follows the same pattern with particle size, but generation 4 is selected since the alpha dose is highest in this region.

6.6 Physical Dose Calculation

The final step in estimating absorbed dose or dose rate to a point within the bronchial epithelium from alpha activity on the airway surface is the application of the physical dose factor (rad per disintegration per cm^2 on the airway surface). The factors reported show remarkable consistency. Four of the published values for RaA and RaC′ on bronchial airways are compared in Figs. 6.5 and 6.6. The factors are calculated as a function of depth below the mucus-tissue interface. The values derived by Haque (1967a) were not given in this form, but were recalculated here as the dose to a 1 μm diameter sphere at various depths so that a comparison with others is possible.

The parameters which have been considered in the derivation of the activity to dose factor are: the airway radius; the location of the activity, either within or upon the mucous blanket; the thickness of the mucus layer; the contribution to dose from the airway surface across the bronchial lumen; and the stopping power or energy loss of the alpha particle in tissue.

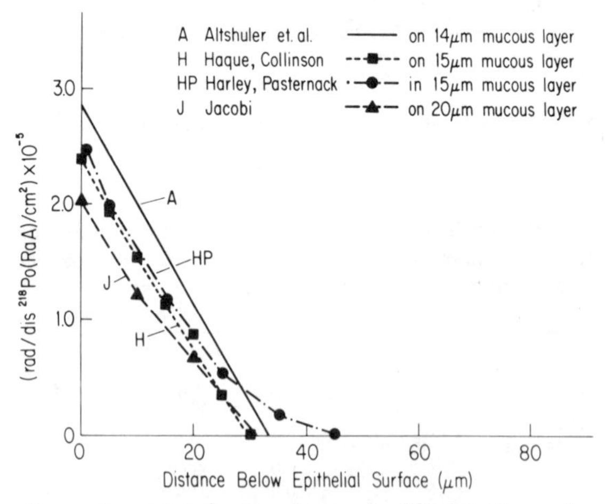

Fig. 6.5. Comparison of alpha dose factors for ^{218}Po(RaA) as a function of depth into bronchial epithelium.

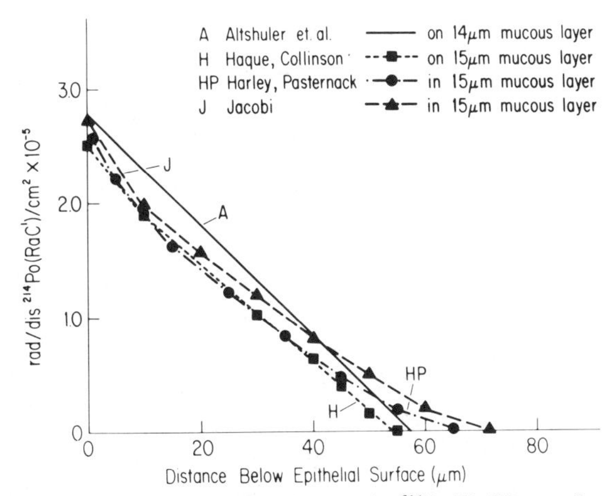

Fig. 6.6. Comparison of alpha dose factors for ^{214}Po(RaC′) as a function of depth into bronchial epithelium.

The total thickness of the total mucous blanket is generally accepted as about 15 μm (7 μm for the serous or sol layer and 7 μm for the viscous or mucous layer) in the normal human adult. In the various calculations, the radon daughters have been assumed to be either on the surface of the mucus or mixed within it. If the daughters are on the mucus, the alpha particle cannot penetrate as far into epithelial tissue since mucus is similar to tissue in its energy absorption characteristics.

It should be more realistic to consider radon daughter activity to be mixed within the mucous layer. It seems more likely that mixing within the total 15 μm blanket is in better accord with functional clearance mechanisms and there is some evidence that this is so (Kirichenco *et al.*, 1970).

The airway radius has been shown to have a small effect on the dose conversion factors. Over the entire depth dose curve, that for the trachea is usually less than 20 percent smaller than that for the terminal bronchioles (0.06 cm diameter) (Haque, 1967a, Harley and Pasternack, 1972).

Comparing the four published values in Figs. 6.5 and 6.6, the values of Altshuler *et al.* (1964) are about 30 percent larger than those of Haque (1967a), Harley and Pasternack (1972), or Jacobi (1964). Altshuler *et al.* (1964) estimated their conversion factor by assuming the daughters to be on the mucous surface and that the dose rate at the surface was twice the average dose rate, decreasing linearly to zero at the range of the alpha particle (47 μm for RaA and 71 μm for RaC′). This approximation ignores the alpha energy lost in air in the lumen and therefore the estimate should be somewhat high. Jacobi (1964) in his estimation procedure calculated the near-wall bronchial dose using stopping power values calculated by Neufeld and Snyder (1961). He then doubled the values for the total dose. That is, near wall plus far wall. This results in an overestimation of the total dose, but the stopping power values that he used are now regarded as somewhat low, which compensated for this overestimation. Jacobi (1964) assumed that RaA was on the mucous surface and RaC′ mixed within it. Jones *et al.* (1978) have also calculated the dose from RaA but only for the near-wall dose, so these values are not plotted in Fig. 6.4. Their near-wall results for RaA mixed in mucus are essentially the same as those of Harley and Pasternack (1982).

Haque (1967a) utilized the measured stopping power data of Riezler and Schepers (1961) and considered it sufficiently accurate to normalize their air data to tissue. Harley and Pasternack (1982) measured the stopping power for the RaA and RaC′ alpha particles in polycarbonate, a polymer with ionization potential near that of tissue. Polynomials fit to the measured data were used in the calculation of their dose conversion factor. Either of these techniques appear to yield similar results.

Thus, regardless of the approach, the spatial dose factors agree well. The error that exists in estimating the dose to a point below the surface of the bronchial epithelium is small unless the site chosen is near either the RaA or RaC′ alpha particle range where the variation due to the choice of activity distribution at the mucous layer becomes significant.

At a point 22 μm below the epithelial surface, the conversion factors are 0.75×10^{-5} and 1.4×10^{-5} rad per alpha disintegration/cm^2 for RaA and RaC′ or 8.8 and 16 rad/year per pCi/cm^2 for RaA and RaC′, respectively.

6.7 Breathing Pattern (Including Nasal Deposition)

Total radon daughter deposition rate in the bronchial tree depends upon the daughter concentration in air, the tidal volume (volume per breath), and the breathing frequency. Tidal volume and breathing frequency determine the flow rate in the trachea, while the flow rate in subsequent generations may be calculated for the Weibel model assuming dichotomy. Eqs. (6-8) and (6-9) may then be used to determine fractional deposition in each airway. Correction should be made in the upper airways for convective diffusion (Table 5.1) and fractional deposition should be related to unit input into the trachea if there is a substantial loss in total activity as in the case of unattached RaA. The total deposition rate in a particular airway, i, is then calculated from:

$$d_i = \mathrm{N}\mathrm{f}_i\, TBC_{\mathrm{n}} \tag{6-11}$$

where

d_i = deposition rate in airway i (pCi/min),
N = penetration fraction,
f_i = fractional deposition in airway i expressed as a fraction of input to the trachea,
T = tidal volume (liters/breath),
B = breathing frequency (breaths/min), and
C_{n} = concentration of radon daughter n (RaA, RaB, RaC) in pCi/L.

Deposition during both inhalation and exhalation must be accounted for.

Although increased breathing frequency for a given tidal volume increases the total radon daughter inhalation rate, the flow rate in the bronchial tree is also increased which decreases fractional deposition [Eq. (6-8)]. These factors are not completely compensating but do tend to keep deposition in each airway similar over a range of breathing patterns. That is, the product (f_i) (T) (B) is reasonably constant. This effect is illustrated in Table 5.4. If the equilibrium values of alpha activity of the attached daughters in Weibel Generation 4 are compared for the active and resting male, for example, a difference of about 30 percent is evident even though the breathing rates differ by a factor of two (18.75 L/min versus 9 L/min).

Nasal deposition enters the deposition Eq. (6-11) as a direct constant. Ordinarily the penetration fraction is considered to be 40 percent to 50 percent for unattached RaA and 98 percent for the attached fraction of the daughters (Altshuler *et al.*, 1964; George *et al.*, 1970). The values in Table 5.4 were calculated using 40 percent and 98 percent, respectively. For mouth breathing, there is no nasal deposition and the equilibrium activity of unattached RaA in Table 5.4 should be multiplied by $1/0.4 = 2.5$. Mouth breathing increases the calculated annual dose by about 35 percent.

The range of annual dose as a function of breathing pattern may be calculated assuming that the dose for the active breathing pattern represents the highest possible dose and that for a resting pattern the lowest. Examination of Eqs. (5-8) to (5-13) shows that the resting pattern would be 25 percent less and the active pattern 25 percent greater than the nominal annual dose.

6.8 Bronchial Morphometry

There are limited data on bronchial morphometry. At the present time there are no detailed data on the bronchial morphometry of either women or children. The values reported by Weibel (1963) are based on the casts of lungs from two males. Measurements of a few casts of the normal bronchial tree have been published by Horsfield *et al.* (1971) and Schlesinger *et al.* (1977). The most extensive measurements so far are those of Raabe *et al.* (1976) and Yeh and Schum (1980) and include not only bronchial morphometry data for individual lobes of the human lung but for that of the dog, the rat, and the hamster. Phalen *et al.* (1978) have summarized these data for modeling purposes.

Radon daughter activity in the human bronchial tree has been calculated mainly for simple dichotomous branching in this report. However, Harley and Pasternack (1982) have utilized the morphometric data of Yeh and Schum (1980) to calculate radon daughter dose conversion factors in the five-lobed human lung. Their values do not differ from those calculated using the simple Weibel dichotomous branching by more than 30 percent depending upon the airway chosen. James *et al.* (1981) have also performed comparisons between the Yeh and Schum and Weibel morphometry with similar results. In the normal lung, individual differences in a single airway could conceivably produce dose rate variations but deviations from average morphometry and simple dichotomy for generation 4 are unknown.

6.9 Mucociliary Clearance

The equilibrium activity established on the bronchial airways is the result of deposition, radioactive build-up and mucociliary clearance of the radon daughter nuclides. Measurement of the mucociliary clearance rate in humans has not been possible so far except for some values obtained in the trachea. Goodman *et al.* (1978) measured tracheal clearance rates of about 1.0 cm/min in young nonsmokers. They obtained values of about one-half this for elderly persons and smokers. Whether this reduction in tracheal clearance is indicative of changes in clearance rates in other airways is not known. In smokers and individuals with impaired clearance due to chronic disease, such as asthma, complete stasis has been observed for periods of up to a few hours in experiments where subjects inhaled radioactively tagged aerosol particles.

The values of mucociliary clearance rate that are used in assessing equilibrium radon daughter activity are usually calculated. Altshuler *et al.* (1964) calculated mucus transit times for the Landahl lung model based on 1.5 cm/min flow in the trachea and constant production of mucus over the entire surface of the bronchial tree. Jacobi (1964) used values of 1.5 cm/min in the trachea, 0.25 cm/min in the bronchi, and 0.5 cm/min in the bronchioli. Harley and Pasternack (1982) used the results of Altshuler *et al.* for the Landahl model and calculated scaled values for the Weibel lung model. The transit times used in the two studies that estimate equilibrium activity in the Weibel model [Harley and Pasternack (1982) and Haque and Collinson (1967)] are shown in Table 6.4.

Although measured data relating to clearance times are scanty, it is possible to compare the results of the equilibrium activity for different mucus transit times that might occur in real situations. The results of these calculations are shown in Table 6.5. The deposition is calculated

TABLE 6.4—*Mucus transit times used in published alpha dose models employing Weibel bronchial morphometry*

Weibel Generation	Airway Length (cm.)	Mucus Transit Times (Minutes) Used in Weibel Model	
		Harley-Pasternack	Haque-Collinson
0	12	8	9.6
1	4.76	6	3.8
2	1.90	8	7.6
3	0.76	3	3.04
4	1.27	14	25.4
5	1.07	12	21.4
6	0.90	10	18
7	0.76	32	Long
8	0.64	27	
9	0.54	23	
≥10	0.46	Long	

TABLE 6.5—*Steady state areal alpha activity (pCi/cm²) in Weibel Generation 4 for different mucus clearance times. Atmospheric concentrations are 1000 pCi/m³ of each inhaled nuclide, unattached RaA, attached RaA, RaB, and RaC. Breathing pattern is that of the active male.*

Inhaled Nuclide	RaA*		RaA		RaB		RaC	
Equilibrium Activity	RaA	RaC′	RaA	RaC′	RaA	RaC′	RaA	RaC′
Mucociliary Clearance Values								
Harley-Pasternack Table 6.4	0.056	0.030	0.0013	0.0011	—	0.0096	—	0.0082
Haque-Collinson Table 6.4	0.056	0.029	0.0013	0.0008	—	0.0071	—	0.0075
Rapid Clearance (Harley-Pasternack x2)	0.054	0.017	0.0013	0.0007	—	0.0069	—	0.0073
Complete Stasis on Entire Tree	0.055	0.055	0.0012	0.0012	—	0.011	—	0.0080
Stasis in Generation 4, Harley-Pasternack Elsewhere	0.072	0.19	0.0017	0.0061	—	0.051	—	0.023

according to Gormley-Kennedy (1969) corrected by Martin and Jacobi (1972) factors and the clearance times are varied. The equilibrium activity for five different mucus transit conditions are compared. These are: mucus transit time calculated by Harley and Pasternack (1972), by Haque and Collinson (1967), by assuming more rapid clearance (a factor of two over the values of Harley and Pasternack), by assuming stasis or no clearance on the whole bronchial tree, and by assuming stasis only in generation 4.

The last assumption represents the worst possible case and the highest alpha dose since the more distal generations bring activity into generation 4 with no clearance from it. This case may be possible physiologically, but certainly only on a transient basis. Even for this worst case it is seen in Table 6.5 that the values of equilibrium activity are about a factor of three higher for unattached RaA and about a factor of 3 to 7 for the attached daughters. The difference in equilibrium activity in airway 4 resulting from the Harley-Pasternack transit times and those of Haque-Collinson is small even though the transit times differ by a factor of almost two in generations 4 to 6. Thus, changes in mucociliary clearance rates do not introduce large differences into the dose models unless real impairment is present.

James (1976) proposed an alternative clearance mechanism for ^{214}Bi(RaC) from the bronchial tree, namely, by diffusion to blood. He measured the transfer to blood of ^{212}Pb and ^{212}Bi for a tracheal instillation of carrier-free solution in rabbits. Transfer of ^{212}Pb to blood indicated that removal of RaB via this route would be negligible. However, 30 percent of the ^{212}Bi was cleared to blood with a half-time of 7 minutes. It is not known whether ^{214}Bi(RaC) is removed from the carrier aerosol so that it might suffer this same transport. If the magnitude and half-time for removal to blood as proposed by James (1976) apply to aserosols deposited in the human bronchial tree then the equilibrium activity of RaC′ arising from either inhaled RaB or inhaled RaC would be 20 percent less than the values calculated in any of the tables in Sections 5 or 6.

6.10 The Location of Target Cells

Mature or differentiated cells of the bronchial epithelium do not divide. They are lost by desquamation into the lumen and replaced by division and differentiation of basal cells (Breeze and Wheeldon, 1977). It has been known since 1881 that mature ciliated and globlet cells in the bronchial epithelium develop from basal cells (Drusch, 1881; Bockendahl, 1885). Basal cells are found closely spaced and in a single layer along the basement membrane of the larger bronchi but they become sparsely distributed in the bronchioles (bronchi with no cartilage as solid support) (Weibel, 1963). Goldziecher (1918) was the first to suggest that small cell carcinoma had its origin in the basal or germinal cell layer. In the early 1960's, it was suggested that these were the target cells in the formation of radiation-induced tumors in uranium miners (Altshuler *et al.*, 1964; Jacobi, 1964). The malignant transformation itself was assumed to take place within the basal cell nucleus (Kotin *et al.*, 1966).

Within the last ten years, another cell which is found near the basal layer has been implicated in the origin of small cell tumors. It was known that carcinoid tumors of the intestinal tract were derived from certain endocrine cells (Kultschitsky cells). Studies of small cell or oat cell carcinoma in the bronchial tree show the presence of cells that closely resemble the Kultschitsky cell and this cell, although given many names, is now commonly called the K-cell. Certain hormonal disturbances in individuals with small cell carcinoma support the K-cell origin in this particular tumor (Feyrter, 1954; Bensch *et al.*, 1968; Hattori *et al.*, 1972). Clusters of K-cells have also been observed in bronchial epithelium and these clusters have been termed neuroepithelial bodies (Laurweryns *et al.*, 1974; Cutz *et al.*, 1975).

The K-cell is present at all levels of the bronchial tree but particularly at bifurcations. It is found only in small numbers in bronchial epithelium and for this reason is probably not implicated in all radiation-induced small cell carcinoma. However, it should be kept in mind, along with the other abundant basal cells, as a possible target.

For some years it was thought that radiation-induced tumors were uniquely of the small or oat cell variety. One study (Horacek *et al.*, 1977) now indicates that, in Czechoslovakian uranium miners, 54 percent of the bronchial tumors were small cell carcinomas and 35 percent were epidermoid tumors. The different tumor types may also suggest different cells of origin. Saccomanno *et al.* (1981) have shown that the histological type of cancer among uranium miners is related to cumulative radiation exposure, age at first radon daughter exposure and latent interval (see Section 8.2).

Unfortunately, information on the location of basal cells is scanty and that on K-cells is almost nonexistent. Some measurements of the depth of basal cell nuclei have been made for the specific purpose of dose calculation (Altshuler *et al.*, 1964; Gastineau *et al.*, 1972). Their data and an estimate of the K-cell depth from Bensch *et al.* (1968) are given in Table 6.6.

If we assume that in the extreme case the location of the target cell is not at the depth of shallow basal cells (22 μm) but at the average

TABLE 6.6.—*Measurements of the location of basal cell and K-cell nuclei. Values are distances in μm below epithelial surface*

Cell Type		Bronchial Airway			Investigators[a]
		Main	Lobar	Segmental	
Basal Cell	Minimum Depth	45	35	22	A
		62	20	—	G
	Median Depth	82	56	49	A
		70	42	42	G
	Maximum Depth	75	60	90	G
K-Cell		—	—	~25[b]	B

[a] A Altshuler *et al.*, 1964
G Gastineau *et al.*, 1972
B Bensch *et al.*, 1968

[b] Specific airway unknown. Total thickness of bronchial epithelium was 40 μm indicating that the specimen was from either lobar or segmental bronchi.

basal cell depth of 45 μm, then the dose will be due entirely to RaC′ (45 μm corresponds to nearly the range of the RaA alpha particle). From Fig. 6.6, the RaC′ dose is reduced from 1.4×10^{-5} to 0.46×10^{-5} rad per disintegration per cm^2 or 70 percent and the annual dose estimate from Eqs. (5-2) to (5-7) is reduced by about 80 percent.

6.11 Mucus Thickness

Lucas and Douglas (1934) first proposed the two-layer model of mucus overlying bronchial epithelium. According to them, the cilia are covered with a sol layer and over this a gel layer moves, carrying particles to the pharynx. Although the sol layer has not been observed directly, the mucous layer has been studied by several investigators, Yoneda (1976), Van As and Webster (1974), Holma (1969), Nowell and Tyler (1971), Sade *et al.* (1970), and Asmundsson and Kilburn (1970). Wanner (1977) has published a detailed summary of mucociliary transport. Ciliated cells contain approximately 200 cilia and their average length in the bronchi is 7.5 μm (Gastineau *et al.*, 1972). The fluid covering the cilia could vary from a polymolecular layer which would serve only as a ciliary lubricant or could immerse the entire length of the cilia (Van As, 1977). Mucus and particles, when in motion, should be transported at the tips of the cilia. Particles can probably be located anywhere within mucus during transport out of the bronchial tree. Van As and Webster (1974) have studied the mucous layer in rats and report that mucus is discontinuous and is present in droplets, flakes, and plaques. Sturgess (1977), on the other hand, claims that scanning electron microscopy of the rabbit lung indicates that mucous secretion does form a continuous layer overlying the major airways.

Thus, it is possible that in the normal lung, particles can be present directly on the bronchial epithelium or as far away as about 15 μm due to the presence of the total mucous blanket (sol plus gel). In smokers or persons with obstructive lung disease, mucus production can be significantly larger than normal and mucus plugging has been reported, for example, in the bronchioles of young cigarette smokers (Niewoehner *et al.*, 1974). In smokers, varying degrees of denudation of the ciliated epithelium have been observed and in such areas it is also possible for particles to reside on the epithelial surface. For abnormally thick mucus, such as with plugging, the mucus may actually have a protective effect and effectively eliminate the probability of an alpha particle reaching cells in bronchial epithelium. Axelson and Sundell (1978) have found that in a small group of Swedish zinc-lead miners there seems to be a higher lifetime risk of lung cancer among nonsmokers than smokers. However, the smokers with lung cancer died approximately 11 years earlier than the nonsmokers. They propose that increased mucus production in smokers and a thicker mucous layer resulting in a lower alpha dose may be a reason for their observations of lifetime risk, but that smoking promotes the development of cancer once induced, which would account for the shorter latent interval.

Although the distribution of radon daughters on the epithelial surface is not known precisely, Figs. 6.5 and 6.6 may be used to estimate the range of possible effects upon alpha dose. Since the alpha dose in mucus is similar to that within tissue, the effect of varying mucus thickness can be estimated by either increasing or decreasing the depth in tissue around the site where dose is to be calculated. If the daughters are assumed to be mixed within the mucus, then removal of the 15 μm mucous blanket brings the dose site about 1/2 this value or 7.5 μm closer to the source. For the site at 22 μm below the epithelial surface, the dose from RaA of 0.75×10^{-5} rad/disintegration$\cdot$cm^2 would then be about equal to the dose at $22 - 7.5 = 14.5$ μm or 1.25×10^{-5} rad/disintegration$\cdot$cm^2, that is 67 percent higher. For RaC$'$, the dose would be increased by only 20 percent. This would increase the annual dose calculated from Equations (5-2) to (5-7) by 22 percent. It is difficult to calculate the effect of increased mucus thickness since this could conceivably reduce the dose almost to zero. However, if we assume that an additional 15 μm layer is introduced, then the dose would be about equal to that at 22 + 7.5 or 29.5 μm. This reduces the dose from RaA by 50 percent and that from RaC$'$ by 25 percent. This would decrease the annual dose calculated from Eqs. (5-2) to (5-7) by 30 percent.

6.12 Variability

For each of the factors which affect bronchial dose that are discussed in this chapter, the effect upon annual dose is estimated and the range is given as a percentage change from the nominal value. The annual bronchial alpha dose is dependent primarily upon these ten variables and a general form of the dose equation is:

$$D = \text{Annual Dose} = f(x_1 \cdot\ x_2 . \ldots x_{10}) \qquad (6\text{-}12)$$

where $x_{1\text{-}10}$ = the factors discussed in Section 6. The exact mathematical form is not known for each of the factors in Eq. (6-12), but over the indicated range of the parameters, each is assumed linear. Most of the factors can be regarded as independent or uncorrelated with the exception of mucus thickness and mucociliary airway clearance time. It is well known that clearance time is dependent upon mucus thickness with optimum particle transit occurring at a critical mucus thickness and slower motion when the layer is thicker or thinner (Blake, 1975; Barnett and Miller, 1966). However, assuming that the variables are uncorrelated and equally weighted in the dose calculation, it is possible to calculate the variance of the annual dose, D, if the variance of each parameter can be determined. The equation for the variance of the random error is:

$$\text{Var}\ (D) = \text{Var}\ (x_1) + \text{Var}\ (x_2) + \ldots \text{Var}\ (x_{10}) \qquad (6\text{-}13)$$

It is possible to make the approximation that:

$$\text{Var}\ (x_i) = (\text{Range}\ x_i)^2/2 \qquad (6\text{-}14)$$

The range of each of the ten factors is summarized in Table 6.7. The range is expressed as a percent change from the nominal annual dose for males (from Section 5, 190 mrad). Excluding mouth breathing as an additional factor, the variance of the dose expressed as a percent is:

$$\text{Var}(D) = (1/2)(40^2 + 20^2 + 120^2 + 45^2 + 60^2 + 50^2 + 10^2 + 80^2) = 1/2(31025)$$

Then, the standard deviation is:

$$S = \sqrt{\text{Var}(D)} = 120 \text{ percent} \qquad (6\text{-}15)$$

That is, the standard deviation of the annual dose to an individual is 120 percent if all of the factors in the dose estimation are unspecified, except the radon concentration. In a particular environment not all of these variables need be unknown; for example, unattached RaA can

TABLE 6.7—*Variability of radon daughter bronchial dose range as a percent change from nominal annual dose for males*

Factor	Variability	
1. Unattached RaA (variable from 4% to 20%)	−10 +30	(4% unattached RaA) (20% unattached RaA)
2. Daughter product equilibrium (1/0.9/0.7/0.7 to 1/0.9/0.6/0.4)	−20	
3. Particle deposition models		a few percent
4. Particle size (change in median size from 0.125 μm to 0.05 μm or 0.17 μm)	+100 −20	(for 0.05 μm particles) (for 0.17 μm particles)
5. Calculated physical dose to a given site in epithelium		a few percent
6. Breathing pattern (entirely active pattern to entirely resting) (mouth versus nose breathing)	+20 −25 (+35)	(active pattern) (resting pattern)
7. Bronchial morphometry (child versus adult)	+60	
8. Mucociliary clearance (normal to complete stasis)	+10	
9. Mucus thickness (none to twice normal)	+20 −30	(no mucus) (twice normal mucus thickness)
10. Location of target cell (shallow −22 μm to average −45 μm).	−80	

be measured, the breathing pattern for the population is reasonably stable, etc. With this additional information, the error in the bronchial dose calculations ought then to be less than 100 percent, that is, better than a factor of two.

6.13 Summary

The overall variability in the radon daughter dose conversion factor can be as high as about 120 percent if all parameters except the ^{222}Rn concentration are unspecified. Ordinarily, the range of these parameters, such as fraction of unattached RaA, daughter disequilibrium, particle size, etc., is not large and the overall variability would be smaller than this maximum.

7. The Adequacy of the Working Level as an Exposure Unit

7.1 Introduction

Occupational exposures, particularly those of uranium miners, have been described in terms of the working level (WL). The WL was defined specifically for occupational exposure as any combination of short-lived radon daughters in one liter of air that will result in the emission of 1.3×10^5 MeV of potential alpha energy. One calculation showing the details of the decay and energy release for both ^{222}Rn and ^{220}Rn daughters appears in the UNSCEAR (1977) report and is reproduced in Table 7.1. The concept of WL was developed in an attempt to simplify measurements of exposure to radon and its daughters when radioactive equilibrium did not exist (Holaday *et al.*, 1957). Also, it gives a direct indication of the energy released in the whole lung which ordinarily would be related to the significant absorbed dose. The WL has been of great utility in that it is easily measured, usually with only a single alpha measurement on a filtered air sample. A serious drawback is that the WL is not readily related to bronchial dose and is thus not readily comparable with other radiation doses received by the body.

Permissible levels for human inhalation exposure have historically been related to the air concentration of the radioactive contaminant. The characteristic which makes alpha emitters and therefore radon daughters unique is that a whole organ (lung) dose does not appear to be related to the incidence of lung cancer. The pulmonary parenchyma is not found to be the site of tumors in underground miners, but rather the cancer site is commonly in specific airways in the upper bronchial tree.

Apparently, only a small fraction of the total alpha energy released in the whole lung is effective, namely, that to a relatively small area of bronchial epithelium. Thus, otherwise neglected parameters which determine deposition in this region such as fraction of unattached

TABLE 7.1—*Decay characteristics and potential energy of* ^{222}Rn *and* ^{220}Rn *daughters (from UNSCEAR, 1977)*

Radionuclide	Half-Life	Atoms per pCi	Potential Alpha Energy		Conversion Factor	
			per Atom	per pCi	MeV/L / pCi/L	WL / pCi/L
^{218}Po (RaA)	3.05 min	9.77	13.68	134	134	0.00103
^{214}Pb (RaB)	26.8 min	85.8	7.68	659	659	0.00507
^{214}Bi (RaC)	19.7 min	63.1	7.68	485	485	0.00373
^{214}Po (RaC′)	1.6×10^{-4} s	1.0×10^{-5}	7.68	7.68×10^{-5}	7.68×10^{-5}	6.0×10^{-10}
^{216}Po (ThA)	0.158 s	0.00844	14.57	0.123	0.123	9.5×10^{-7}
^{212}Pb (ThB)	10.6 h	2040	7.79	15900	15900	0.1223
^{212}Bi (ThC)	60.5 min	194	7.79	1510	1510	0.0116
^{212}Po (ThC′)	3×10^{-7} s	1.6×10^{-8}	8.78	1.4×10^{-7}	1.4×10^{-7}	1.0×10^{-12}

RaA, particle size, breathing rate, etc. become significant. Each of these factors and its effect upon bronchial dose is discussed in detail in Section 6.

A dosimetric approach for establishing risk is important since this allows direct comparison with response to other types of radiation. However, it is unlikely that a direct measurement of absorbed alpha dose in bronchial epithelium can ever be obtained in humans for radon daughters, and calculations must suffice. Therefore, measurement of the atmospheric characteristics is important because these factors become the basis of the dose estimation.

7.2 Exposure in Working Levels Versus Absorbed Tissue Dose

After a substantial amount of data has been collected in widely different environments, the various factors of unattached RaA, particle size, and daughter ratio may be found to vary only within narrow limits. If this is the case, then the WL may be an adequate unit to describe dose response since a single exposure-to-dose conversion factor (rad/y per WL) would then apply. Individual dose variability would then reside primarily in the biological parameters such as lung size and structure and breathing patterns.

However, until such data become available, the difficulty with WL as a substitute for absorbed bronchial dose may be addressed by comparing any one of the dose equations in Section 5 (Eqs. 5.2 to 5.7) with that for working level.

$$\mathrm{WL} = 1.03 \times 10^{-6}(\mathrm{RaA}) + 5.07 \times 10^{-6}(\mathrm{RaB}) + 3.73 \times 10^{-6}(\mathrm{RaC}) \qquad (7\text{-}1)$$

where RaA, RaB, RaC = atmospheric concentration of total RaA, (Attached + Unattached), RaB, and RaC in $\mathrm{pCi/m^3}$.

The Eq. (5-2) for absorbed alpha dose rate to males with a breathing pattern representing light activity is:

$$\text{Annual alpha dose (mrad/year)} = 0.98(\mathrm{RaA^*}) + 0.029(\mathrm{RaA}) + 0.16(\mathrm{RaB}) + 0.14(\mathrm{RaC})$$

where RaA* = atmospheric concentration of the unattached fraction of RaA in $\mathrm{pCi/m^3}$.

The absorbed dose here is, as usual, that to Weibel generation 4 at 22 μm below the surface of the bronchial epithelium.

A comparison of these two expressions indicates that no provision is made in the WL expression to accommodate the unattached RaA. Also, absorbed dose differences due to breathing patterns and bronchial morphometry, as described by the dose Eqs. (5-3) to (5-7) are not expressed through the WL value.

The similarity in the ratios of the coefficients for the attached fraction of daughters (the last three terms of 5-2) with those of the working level Eq. (7-1), leads to the approximation that for continuous environmental exposure:

$$\text{Annual dose (male, light activity) mrad/year} \cong (\text{RaA}^*) + 32000\ (\text{WL}) \quad (7\text{-}2)$$

Although Eq. (7-2) is an approximation, it yields a numerical result which is within 10 percent of the more detailed calculation for adult males (Eq. 5-2). However, it must be remembered that a variety of modified expressions exists to evaluate resting versus active breathing patterns for adult male, adult female, and child.

It must be emphasized that each approximation smoothes the details of any population dose distribution and should be avoided until it is shown that these details do not change the annual dose substantially.

7.3 Cumulative Exposure (Working Level Months)

One additional problem arises with the calculation of the time-integrated exposure to radon daughters used to assess occupational exposure. Exposure to an atmospheric concentration of one WL for a working month (170 hours) is defined as a cumulative exposure of one working level month (WLM). The total exposure in WLM may be evaluated:

$$\text{Cumulative Exposure in WLM} = \sum_{i=1}^{n} (\text{WL})_i \left(\frac{t_i}{170}\right) \quad (7\text{-}3)$$

where

$(\text{WL})_i$ = average concentration of radon daughters during exposure interval i, expressed in WL.

t_i = number of hours duration of exposure i.

Environmental as well as occupational exposures are now commonly expressed in terms of WLM. The cumulative exposure at a given radon daughter concentration in WL on a continuous basis is four times that for occupational exposure (8766 versus 2000 hours, for an annual exposure). Although Eq. (7-3) is proper for calculating environmental or occupational exposure, this indicates the awkward properties of the WLM unit.

7.4 Absorbed Dose per WLM

The main problem is identified if the absorbed dose to the bronchial epithelium per WLM is calculated for the various occupational and environmental exposures discussed in this report. The median values for measured WL associated with the measured ^{222}Rn concentrations described in Section 5 of 830 pCi/m^3 indoors and 180 pCi/m^3 outdoors are 0.0059 and 0.0013, respectively (George and Breslin, 1980). If their approximation of 6×10^{-6} per pCi $^{222}Rn/m^3$ had been used, 0.0050 and 0.0011 WL would have been calculated. From Eq. (7-3), adults spending 4 hours per day outdoors and 20 hours per day indoors in this environment accumulate an exposure of:

$$\text{WLM per year} = (0.0059)(20 \times 365/170) + (0.0013)(4 \times 365/170) = 0.27 \quad (7\text{-}4)$$

and a ten-year-old child spending 6 hours per day outdoors and 18 hours per day indoors accumulates

$$\text{WLM per year} = (0.0059)(18 \times 365/170) + (0.0013)(6 \times 365/170) = 0.24 \quad (7\text{-}5)$$

The average annual bronchial dose estimates derived in Section 5 [Eqs. (5-17) to (5-19)] for these conditions are 190 mrad for adult males, 170 mrad for adult females, and 300 mrad for a ten-year-old child. The ratios of absorbed radiation dose to exposure in rad/WLM for these individuals are:

$$\text{Adult Male: } 0.19 \text{ rad y}^{-1}/0.27 \text{ WLM y}^{-1} = 0.71 \text{ rad/WLM} \quad (7\text{-}6)$$

$$\text{Adult Female: } 0.17 \text{ rad y}^{-1}/0.27 \text{ WLM y}^{-1} = 0.63 \text{ rad/WLM} \quad (7\text{-}7)$$

$$\text{Ten-year-old Child: } 0.30 \text{ rad y}^{-1}/0.25 \text{ WLM y}^{-1} = 1.25 \text{ rad/WLM} \quad (7\text{-}8)$$

These may be compared with the average value derived for underground uranium miners. For their case, unattached RaA has been found to be lower than normal environmental conditions (0.04 versus 0.07) and their breathing rates are higher during heavy work (see

Table 7.4):

Uranium Miner ~0.5 rad/WLM.

7.5 Adequacy of the WLM Standard in Uranium Mining

A further question needing resolution is whether the WLM standard is adequate for uranium mining conditions. It has sometimes been stated that, if the hazard is attributed to the unattached fraction of the activity, the total alpha energy is not at all representative of the risk, whereas the radon concentration *is* indicative (Pradel, 1973). Inherent in this statement is the possibility of the greater suitability of a dosimetric standard (ICRP, 1982) over the fixed 4 WLM per year exposure standard, because the unattached fraction is taken into account in the former and ignored in the latter. Two standards (the ICRP, 1959 recommendation and the 4 WLM per year value in the U.S.) have been contrasted (Cross, 1979). The present ICRP (1981) standard is based upon both dosimetric and epidemiologic considerations.

To determine the contribution of the unattached activity as a fraction of the total dose, the dose is calculated for an equilibrium mixture of daughters and for an atmosphere containing only RaA. The dose is calculated for two different particle sizes, 0.07 μm ($D = 1.3 \times 10^{-5}$ cm^2/s) and 0.3 μm ($D = 1.2 \times 10^{-6}$ cm^2/s) and for nose and mouth breathing.

Tables 7.2, 7.3, 7.4 and 7.5 demonstrate that for nuclei characterized by a mean diffusion coefficient of 1.3×10^{-5} cm^2/s (0.07 μm diameter),

TABLE 7.2—*Dose conversion factor*[a] *and percent contribution for equilibrium mixture of daughters; nuclei 0.07 μm ($D = 1.3 \times 10^{-5}$ cm^2/s), 4% ions ($D = 5.4 \times 10^{-2}$ cm^2/s)*

Generation	Mouth Breathing	Nose Breathing
	rad/WLM	rad/WLM
0–16	1.2 (3% ions, 97% nuclei)[b]	1.1 (1% ions, 99% nuclei)
2–9	1.5 (22% ions, 78% nuclei)	1.3 (10% ions, 90% nuclei)
4–6	0.98 (30% ions, 70% nuclei)	0.80 (15% ions, 85% nuclei)
7	1.9 (24% ions, 76% nuclei)	1.6 (12% ions, 88% nuclei)
4	1.7 (28% ions, 72% nuclei)	1.4 (14% ions, 86% nuclei)

[a] Tables 7.2 to 7.5 were calculated using the "updated" Haque and Collinson model. For nose breathing, nasal penetration fractions of 0.40 for free ions and 0.0987 for attached daughters or nuclei are used. Breathing rate is for uranium miners, 20 L/min.

[b] Values in parentheses indicate percentage of alpha dose attributable to unattached and attached radon daughters (ions, nuclei).

TABLE 7.3—*Dose conversion factor*[a] *and percent contribution for equilibrium mixture of daughters; nuclei 0.3 μm (D = 1.2 × 10^{-6} cm²/s), 4% ions (D = 5.4 × 10^{-2} cm²/s)*

Generation	Mouth Breathing rad/WLM	Nose Breathing rad/WLM
0–16	0.27 (14% ions, 86% nuclei)[b]	0.24 (6% ions, 94% nuclei)
2–9	0.57 (58% ions, 42% nuclei)	0.37 (35% ions, 64% nuclei)
4–6	0.44 (68% ions, 32% nuclei)	0.26 (46% ions, 54% nuclei)
7	0.74 (61% ions, 39% nuclei)	0.46 (39% ions, 61% nuclei)
4	0.72 (66% ions, 34% nuclei)	0.43 (44% ions, 56% nuclei)

[a] See Table 7.2

[b] Values in parentheses indicate percentage of alpha dose attributable to unattached and attached radon daughters (ions, nuclei)

TABLE 7.4—*Dose conversion factor*[a] *and percent contribution for extreme mine conditions (RaA only); nuclei 0.07 μm (D = 1.3 × 10^{-5} cm²/s), 4% ions (5.4 × 10^{-2} cm²/s)*

Generation	Mouth Breathing rad/WLM	Nose Breathing rad/WLM
0–16	1.3 (26% ions, 74% nuclei)[b]	1.1 (13% ions, 87% nuclei)
2–9	4.2 (76% ions, 24% nuclei)	2.3 (56% ions, 44% nuclei)
4–6	3.5 (82% ions, 18% nuclei)	1.8 (65% ions, 35% nuclei)
7	5.5 (79% ions, 21% nuclei)	2.9 (60% ions, 40% nuclei)
4	5.5 (83% ions, 17% nuclei)	2.7 (66% ions, 34% nuclei)

[a] See Table 7.2

[b] Values in parentheses indicate percentage of alpha dose attributable to unattached and attached radon daughters (ions, nuclei)

TABLE 7.5—*Dose conversion factor*[a] *and percent contribution for extreme mine conditions (RaA only); nuclei 0.3 μm (D = 1.2 × 10^{-6} cm²/s), 4% ions (5.4 × 10^{-2} cm²/s)*

Generation	Mouth Breathing rad/WLM	Nose Breathing rad/WLM
0–16	0.55 (64% ions, 36% nuclei)[b]	0.34 (41% ions, 59% nuclei)
2–9	3.4 (94% ions, 6% nuclei)	1.5 (86% ions, 14% nuclei)
4–6	3.0 (96% ions, 4% nuclei)	1.3 (90% ions, 10% nuclei)
7	4.6 (95% ions, 5% nuclei)	2.0 (88% ions, 12% nuclei)
4	4.8 (96% ions, 4% nuclei)	2.0 (91% ions, 9% nuclei)

[a] See Table 7.2

[b] Values in parentheses indicate percentage of alpha dose attributable to unattached and attached radon daughters (ions, nuclei)

ter), the contribution to the dose by unattached activity is never so large, under equilibrium conditions, as to govern the dose. Total potential alpha energy (WL) appears to adequately correlate exposure and dose. This is especially true for nose breathing. Under conditions of extreme disequilibrium, however, the contribution to the dose of the unattached activity is generally larger and governs the dose, although this is less true under conditions of nose breathing. Harley and Pasternack (1972) state that mouth breathing among uranium miners is uncommon.

From the data in Tables 7.2 to 7.5 then, we conclude that the WL, per se, is a questionable unit for risk assessment unless some allowance is made for the unattached fraction in the WL definition. It is also necessary to determine whether such "extreme mine" conditions represent only a rare, if not nonexistent, exposure condition. In view of other measurements that indicate that average exposure conditions are more like those described by Holleman *et al.* (1968) (1/0.53/0.35) in the Colorado Plateau, the important point appears to be only that the WL unit fails to relate exposure and dose under the extreme conditions of RaA only. It is of further interest to note that, as regards the average dose to the tracheobronchial tree (generation 0-16), the dose conversion factor is relatively constant and the nuclei contribution to the dose is always greater for nose breathing. The tables also indicate that when the dose is averaged over smaller regions of the tracheobronchial tree, the dose conversion factor varies appreciably with the degree of disequilibrium, although the degree of unattachment is of still greater effect. Disequilibrium is treated further in the paper contrasting the ICRP (1959) and WLM exposure standards (Cross, 1979) and in Section 6 of this report.

7.6 Recommendations

The epidemiological data derived from many types of undergound mining show a relatively consistent relationship between lung cancer incidence and exposure to radon daughters in WLM (see Section 8). This indicates that a single average factor (rad/WLM) yields a reasonable first approximation for the pertinent bronchial dose. Consistency in the nature of the mining atmospheres and in breathing patterns would be expected to yield a fairly constant relationship. On the other hand, it is desirable to attempt a more detailed dose estimation if future populations exposed under different conditions are to be studied.

For these reasons, it is recommended that future measurements provide data to allow calculation of a more specific bronchial dose. These measurements should include the concentrations of RaA, RaB,

RaC, and unattached RaA. If unattached RaA cannot be measured, at least the particle concentration (condensation nuclei) should be determined so that unattached RaA may be estimated by the methods described in Section 6.

8. Lung Cancer in Man Attributable to Radon Daughters

8.1 Introduction

The majority of radiogenic lung cancers that have been recorded, have occurred in miners employed where uranium was a significant or predominant mineral. The mines in Joachimstal and Schneeburg in Central Europe began to be worked about 1400 A.D., first for copper, iron and silver, then for several other metals and finally for uranium. A lung disease peculiar to workers in these mines was described as early as 1500 A.D., but was not recognized as cancer until 1879. The etiological role of radon was not suspected until 1932 and not generally accepted until the 1960's (Woodard, 1980).

In 1949, the Atomic Energy Commission, its Grand Junction Operation Office and the Colorado State Department of Health, aware of the European experience, had obtained radon data in several U.S. mines from the Medical Division of the New York Operations Office of the Atomic Energy Commission. The concentrations were high enough so that they appointed a group from various Federal and State Health Agencies to study potential health problems among uranium miners. A prospective study of underground miners was initiated in 1954 (Holaday, 1964), using miners who had been examined in the period 1950–1954.

Bale (1951) and Harley (1952) were the first to point out that the lung cancer hazard was from the alpha dose delivered through lung deposition of the short-lived alpha-emitting daughters of radon and not from the radon itself.

The permissible level for occupational exposure was originally set in 1955 at 10 pCi ^{222}Rn/L by ICRP (1955) and at 100 pCi ^{222}Rn/L by the U.S. Atomic Energy Commission. In 1955, the permissible level of 100 pCi ^{222}Rn per liter in equilibrium with its daughters was proposed in the U.S. along with its equivalent of one working level, which did not require equilibrium (USPHS, 1957). In 1971, based on the preliminary results of the prospective studies of underground miners (Lundin *et al.*, 1971), and recommendations of the Federal Radiation Council, the U.S. Department of the Interior reduced the U.S. occupational standard to four working level months per year with maximum airborne concentrations not to exceed one WL (FR, 1971). This standard is still in effect in the United States.

Lung cancer deaths in excess of those expected (based upon comparison with comparable geographic, age and sex-specific populations) have been observed in epidemiological studies of seven underground miner populations. These are the U.S., Canadian and Czechoslovakian uranium mines, Swedish and British iron miners, Swedish lead and zinc miners, and Newfoundland fluorspar miners. Although other potential carcinogens such as diesel smoke, traces of arsenic or nickel and iron ore are found in these mines, the lung cancer response appears to be correlated with radon daughter exposure. Myers and Stewart (1979) have shown an underlying lung cancer incidence in uranium miners that is not exposure dependent and this may represent the effect of other carcinogens.

All of the studies reported so far suffer from defects. Primarily, exposure conditions are not well known and in some instances the follow-up time for miners is too short.

Periodic measurements of radon and radon daughters have been made in these mines only in the last 30 years (beginning in different mines between 1948 and 1969). However, estimates based upon changes in ventilation and other mining parameters have been made to attempt to estimate exposures reliably. Biases in the reporting of measurements are likely to be present in much of the data. For example, measurements reported by mine operators are likely to be lower than the true average value when there are penalties for having elevated levels (Slade, 1979). By contrast, measurements reported by official mine inspectors are likely to be higher than the true value because they are primarily looking for violations of standards: they will tend to take measurements only in areas which are most likely to be out of compliance. Measurements taken for scientific purposes are best, but they have been so infrequent that few human data can be correlated with them. Because of the probable biases in measurement data, the different mining groups are considered together rather than separately so the measurement biases will have an opportunity to balance each other.

In view of the deficiencies in the data, it is not surprising that the range of reported attributable lung cancer rate (the definition of risk coefficient) for all populations is about a factor of 15 (2.7 to 45 per million per year per WLM). In the lowest exposure category for U.S.

miners (60 WLM) a deficit (statistically not significant) rather than an excess in lung cancers is observed (NAS, 1980).

In this section, the reported human lung cancer experience is reviewed. The basis for selecting a rounded annual rate of 10 lung cancers per million persons per year per WLM based on existing human evidence is established, since an annual risk rate is necessary in estimating the lifetime lung cancer risk from radon daughter exposure.

8.2 Histologic Type of Bronchogenic Lung Cancer Related to Radon Daughter Exposure

It would simplify the epidemiology if the lung cancers induced by radon daughters were different from those occurring spontaneously or those induced by smoking. It does appear that the distribution of cancer types is different although no new type appears.

Following the early reports of excess lung cancer in the U.S. underground uranium miners, one study showed that the bronchogenic carcinomas were primarily of the small cell undifferentiated type (WHO Classification 2A, 2B) (Saccomanno *et al.*, 1964, 1971). Later, as more material became available for histological analysis, it became clear that the number of both epidermoid (WHO Classification 1A, 1B, 1C) and small cell undifferentiated types increase with increasing radon daughter exposure over those expected (Archer *et al.*, 1974). The observed to expected ratio for both these tumor types has now been reported to increase with cumulative radon daughter exposure in the Czechoslovakian uranium miners and in Canadian uranium miners (Horacek *et al.*, 1977; Kunz *et al.*, 1979; Chovil, 1981).

Although it was originally reported that the percentage of small cell carcinomas in the total observed cases could be dependent upon cumulative exposure, this has been shown not to be so (Saccomanno *et al.*, 1971; Archer *et al.*, 1974; Saccomanno *et al.*, 1981). One difficulty in calculating the percentage of small cell carcinomas in a group of supposedly radon daughter induced cancers of all types, is properly accounting for expected tumors in the group based upon smoking history.

It is known that in nonmining smokers about 17 percent of the tumors are small cell undifferentiated tumors (Saccomanno, 1981).

Horacek *et al.*, (1977) report 54 percent small cell undifferentiated tumors (and 35 percent epidermoid) in the Czechoslovakian uranium miners but suggest that this percentage may be affected by the time course of the exposure.

Chovil (1981) reports 52 percent small cell undifferentiated (and 41 percent epidermoid) in 91 Canadian uranium miners. His results, however, were not stated to be corrected for expected lung cancers and so the percentage of small cell carcinomas may be artificially low having the same "dilution" effect first suggested by Saccomanno *et al.* (1971), that is, not accounting for smoking-induced tumors in the radon daughter exposed cohort. Saccomanno (1981) reports that the frequency of small cell undifferentiated tumors in the U.S. mining population uncorrected for smoking history is currently 22 percent, reflecting a dilution effect of expected smoking related cancers (17 percent small cell in nonmining smokers) and a change in frequency over time.

The percentage of small cell undifferentiated tumors occurring in underground hematite miners and fluorspar miners has also been reported. Again, values are not corrected for the expected lung tumors. Boyd *et al.* (1970) reported 37 percent among hematite miners in Cumberland, England and Roussel *et al.* (1964), 44 percent in French hematite miners.

One contrasting study is that of Newfoundland fluorspar miners. Reported radon daughter concentrations in these mines were extremely high, averaging from 2.5 to 10 WL with individual cumulative exposures up to 3000 WLM. Average exposure, however, is estimated as 204 WLM (NAS, 1980). The miners are reported to have 7 percent small cell undifferentiated tumors and 90 percent epidermoid (Wright and Couves, 1977).

Saccomanno *et al.* (1981) report that in 361 cases of lung cancer in U.S. uranium miners, uncorrected for expected cancers, the percentage of epidermoid bronchogenic carcinoma increases with increasing latent period (from 0 to 40 percent for 7 year versus 30 year latent period). A longer latent interval may explain the high percentage of epidermoid carcinoma in Newfoundland miners. Saccomanno *et al.* reported that the percentage of small cell carcinoma decreased from 85 to 40 percent over the same latent period. In their study, the percentage of epidermoid carcinoma also increased with pack years of cigarette smoking. The percentage of both epidermoid and small cell carcinomas was relatively constant as a function of age at start of mining. Adenocarcinoma, however, was age dependent totaling 10 percent of the bronchogenic carcinoma yield if age at start of mining was 16 years versus 0 percent if mining commenced at age 50.

Fig. 8.1 indicates the change in histological type with increase in latent period.

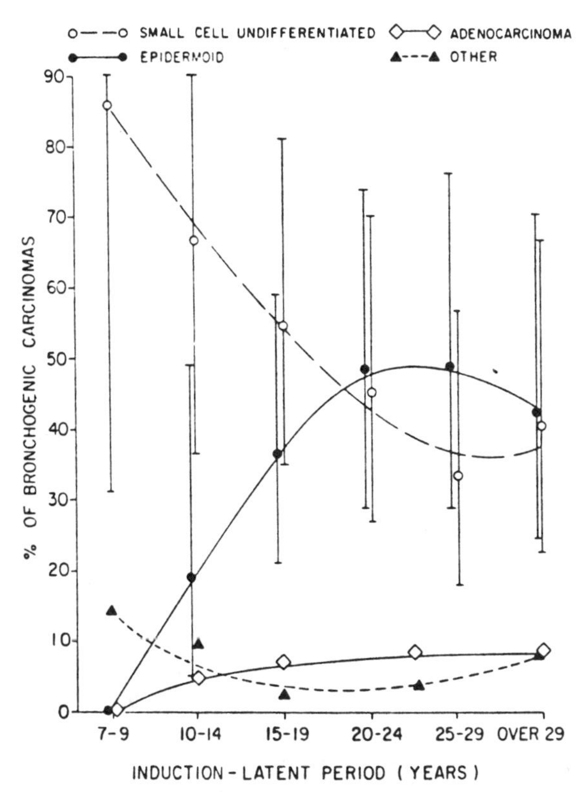

Fig. 8.1. Frequency of histologic types among uranium miners as related to induction-latent period (7+ years latency, 1000 + WLM; 190 cases) (from Saccamanno *et al.*, 1981).

It appears that cell type is not a useful method of differentiating cancers attributable to radon daughters from those occurring in the general population, unless specific information concerning individual exposure and smoking history is available.

8.3 Epidemiological Studies

The assessment of the risk of attributable lung cancer through human epidemiological studies is difficult since the detailed information required is not available. In the ideal case, the exposure of each miner as a function of time would be available and the follow-up period would be sufficient for all of the group to have died from lung cancer or other causes. In addition, it would be possible to separate attributable lung cancers from those arising spontaneously or from cigarette smoking. The cumulative exposure, person years at risk and the number of lung cancers attributable to mining would allow the exact calculation of risk.

The present data do not fulfill these requirements since exposures are only estimates and the follow-up periods are not sufficiently long. Nevertheless, recognizing the limitations of the data, it is possible to estimate a mean risk which is consistent with experience.

8.3.1 *United States Miners*

The uranium miner cohort was assembled between 1950 and 1963 (Lundin *et al.*, 1971; Archer *et al.*, 1973, 1976, 1978, 1979, 1981). It consisted of 3362 white miners and 780 non-white miners (mostly Navajo Indians) who had at least one month of underground employment prior to December 31, 1963. A physical examination and an occupational history was obtained for each miner. Exposure estimates were determined through annual follow-up. Mortality was determined from the period of first examination to September 30, 1974 (Archer *et al.*, 1978). Expected lung cancer rates were calculated on the basis of age and year-specific rates for white males in the states of Colorado, Utah, Arizona, and New Mexico. Expected rates were increased by 10 percent to account for the inclusion of lung cancer cases still living (NAS, 1980). The data cases and person-years accumulated 10 or more years after the start of uranium mining as reported in NAS (1980) are given in Table 8.1. The follow-up periods differ for the different exposure categories. Those in the highest exposure categories started mining before 1955, while most of those in the lowest exposure categories began mining in the 1956–1960 period. There was a scarcity of young miners during 1941–1945 because of the wartime effort. The majority of uranium miners during 1942–1955 were middle-aged men with experience in other types of hard rock mines. Some of them had mined the same ores for vanadium during the 1930–1942 period. Estimates of the radon daughter exposure during this earlier period were included in their exposure history. There was a rather rapid turnover in personnel. The mean time worked underground for those who developed lung cancer was 9.1 years and the mean time from cessation of mining to lung cancer was 10.0 years.

Approximately 43,000 radon daughter measurements were performed between 1951 and 1969, the period during which most of the exposure occurred (Lundin *et al.*, 1971). Most of these measurements were made by official mine inspectors. An estimated 2500 different mines operated during this period, and measurements could not be made in all mines. Interpolation, using computer estimates based on later measurements in the mine or from measurements made in nearby mines was used to estimate exposure in years when no measurements were available and to extrapolate to earlier periods (Lundin *et al.*, 1971).

Among the miner cohort, 54 percent smoked 20 cigarettes/day or fewer, 17 percent smoked more than 20/day, and 29 percent did not smoke cigarettes. This smoking rate was somewhat higher than for the general population of adult males.

The calculated excess attributable lung cancer rate ranges from 2.7×10^{-6} to 8×10^{-6} per person per year per WLM.

8.3.2 *Czechoslovakian Miners*

The mortality in Czechoslovakian uranium miners who started mining from January 1948 through December 1957 has been reported by Sevc *et al.* (1978), Kunz *et al.* (1978) and Kunz *et al.* (1979). The miners were assembled into Group A, those who started mining between January 1, 1948 and December 31, 1952, and Group B, those who started mining between January 1, 1953 and December 31, 1957. Seven exposure categories were reported. The original cut-off date for evaluation of data was December 31, 1973. Detailed follow-up data are not available for Group B, so they are not considered further. Although the number of observed cancers and the exact size of the groups were not given in the initial reports, it was stated that the total group was somewhat larger than the U.S. Study Group (3362 men) and that Group A was 56 percent of the total.

They further subdivided the miners in Group A into three groups, those up to 29 years old, those aged 30–39, and those 40+ years old at start of exposure.

Following their initial publication in 1976, Kunz *et al.* (1978) re-evaluated excess or attributable lung cancers for Group A, using modified life table techniques to make their data more comparable with U.S. data. In this report, exposure categories were condensed to four. They reported that in Group A there were 56,955 person years up to the initial cut off date of December 31, 1973. Kunz *et al.* (1978) report an average observation period of 24.5 years but Sevc (1981) indicates that an observation period of 23.5 years is more precise. This yields about 2400 persons in Group A and thus, about 1900 persons in Group B. It is possible to calculate from the reported observed and expected lung cancer rates per person year (Kunz *et al.*, 1978) that a total of 212 lung cancers were observed in Group A, whereas about 40 were expected or about 170 excess for follow-up through December 31,

TABLE 8.1—*Summary of reported data on lung cancer from five different mining areas*

Source	WLM Category[a]	Person years at risk	Time from start of exposure to cut-off date (y)[b]	Lung cancers observed	Lung cancers expected[c]	Observed cases/y per 10,000	Expected cases/y per 10,000	Reported attributable lung cancers/WLM/ 10^6 persons	Calculated attributable lung cancers/WLM/ y/10^6 persons
U.S.									
NAS (1980)									
U.S. Uranium Miners (10 or	60	5183	16	3	3.96				−3.1
more years after start of min-	180	3308	17	7	2.24				8.0
ing; follow up through Sep-	300	2891	18	9	2.24				7.8
tember, 1974)	480	4171	19	19	3.33				7.8
	720	3294		9	2.62				2.7
	1320	6591	21	40	5.38				4.0
	2760	5690	23	49	4.56				2.8
	7000	1068	24	23	0.91				3.0
Czechoslovakia									
Sevc *et al.* (1976), Kunz *et al.*	72	9380[f]	23.5[e]	6[g]	3.1[g]			100(580)	4.6(36.2)[e]
(1978) Czechoslova-	124	16131	23.5[e]					180(540)	11.2(33.8)
kian Uranium Miners	174		23.5[e]	40	12.3			220(720)	13.8(45.0)
(Group A)	242	19614	23.5[e]					190(590)	11.9(36.9)
	343		23.5[e]	84	15.1			250(340)	15.6(21.2)
	488	11830	23.5[e]					290(250)	22.6(15.6)
	716		23.5[e]	82	10.0			220(160)	17.2(10.0)
Sweden									
a. Jorgensen (1973)									
Kiruna, Iron Miners	34[d]	32900	15	13	4.47				7
b. Axelson (1978)									
Zinkgruven 50–59 yrs	240	1718	39	13	1.3				28.4
70+ yrs	390	437	48	7	1.0				35.0
c. Radford and Reynard (1974)									
Radford (1981)									
Malmberget, Iron Miners	85[i]	13332	16	50	10.1				22.3(16.3)[h]
Canada									
a. Hewitt (1976, 1979) Mc-	21		14	40		4.8	4.0		3.8
Cullough (1979) Canadian	72		15	16		11.9	5.0		9.6
Uranium Miners	180		16	25		28.0	7.0		11.1

Newfoundland						
Devilliers and Windish (1964), Wright and Couves (1977), McCullough *et al.* (1978) NAS (1980), Morrison (1981) Fluorspar Miners	204[j]	28786	29	65 (to 1971)	10	~18
					$\bar{x} \pm S\bar{x}$ (Average of all values except Czechoslovakian values in parentheses)	12 ± 2

[a] Most of these categories represent the median of the interval; a few are mean values. There is some estimation in all groups and some men in all groups started mining before measurements were made.

[b] These values are all estimated. They attempt to represent a mean value for persons in each group.

[c] The expected values for U.S. miners are most accurate. Others involve estimates or indirect calculation.

[d] WLM estimates from ^{222}Rn measurements on the basis 1 WL = 100 pCi ^{222}Rn per liter.

[e] Values for Czechoslovakian miners are for those 30–39 years old at start of exposure. Values in parentheses are for miners 40+ years old at start of exposure. Attributable lung cancers per WLM per year 10^6 persons are calculated for 16 years duration for risk expression. Attributable risk in 72, 488, 716 WLM exposure categories and average years of follow up for Group A, 23.5 from Sevc (1981) (Private Communication).

[f] Person years from Kunz, Sevc and Placek (1978).

[g] Observed and expected lung cancers as of December 31, 1973, calculated from Kunz *et al.* (1978) in combined exposure categories.

[h] Value calculated for smokers, nonsmokers attributable risk in parentheses (Radford, 1981).

[i] Omitting last five years from exposure category as "wasted exposure" yields 85 WLM. Including these last five years yields 97 WLM. Both ^{222}Rn and WL measured.

[j] Exposure estimate from NAS (1980). Attributable lung cancers/WLM·y·10^6 persons from NAS (1980).

1973. In a study of mortality experience in Group A as a function of length of exposure (5.6, 9.5, or 14 years duration), Kunz *et al.* (1979) subsequently reported 80.9 total attributable lung cancers per 1000 miners followed through December 31, 1975. This would yield about 200 excess lung cancers in Group A as of the later cut-off date.

The attributable risk for the entire Group A (including all three age subgroups) is 11×10^{-6} per year per WLM. This value was calculated from the attributable rates and person-years at risk (PYR) reported by Kunz *et al.* (1978). Their reported PYR, however, does not account for any latent interval. Also, it includes the younger age subgroup (29 years at start of exposure) and so includes PYR before age 40 when lung cancer is not expected to occur. Both these factors tend to reduce the overall attributable risk rate. Attributable risk for miners from Group A who were 30 to 39 years or 40+ years old at start of mining are also shown in Table 8.1. Because lung cancer rarely appears before age 40, the age 30 or older subgroups probably reflect more representative values of attributable risk. Follow-up of the under 29-year-old group is needed to complete the picture. Up until 1973 the attributable lung cancer rate for the entire under 29-year subgroup was about 60 percent of the value for the 30–39 year subgroup, and individual exposure categories differed by up to a factor of two.

Over 120,000 radon measurements were performed after 1948. However, an estimate of working level had to be made since radon daughters were not measured. It is not reported who made the mine measurements, but they were presumably made under the direction of mine management. The authors believed that their exposure data were better than those used for the U.S. Study Group.

Cigarette smoking habits were investigated in a randomly selected group of 700 uranium miners. About 70 percent of them smoked cigarettes, just about the same as in the general population of males in Czechoslovakia. Therefore, expected lung cancer values were calculated for each year for each exposure group on the basis of age-specific mortality rates for the male population of Czechoslovakia, and were added to give the total expected number of lung cancer deaths. The attributable cases (observed minus expected) were given for Group A stratified by WLM and age at start of mining. Statistically significant differences betwen observed and expected were found only above 100 WLM cumulative exposure.

Observation of the U.S. uranium miners and the older data from Schneeburg and Joachimsthal (Hueper, 1942) indicate that radon daughter cancer rarely, if ever, appears within seven years after start of exposure. It is therefore assumed that all excess cancers occurred later than seven years after start of exposure. Accordingly, the observed excess per WLM per million persons reported by Sevc *et al.* (1976) has been divided here by 16 instead of the 23.5 average years of follow-up to obtain the excess cases per year per million persons. The attributable lung cancer risk calculated in this manner for both the 30 to 39 year, and the 40+ year (age at start of exposure) subgroups is given in the last column in Table 8.1. The average calculated attributable risk rate for the 30 to 39 year group is 13.8×10^{-6} per person per year per WLM [close to that for the entire group assuming no latency (11×10^{-6})] and that for the 40+ group is 28×10^{-6} per person per year per WLM.

The older age group (40+ at start of mining) has a statistically higher attributable risk rate in the 174 and 242 WLM categories. This may reflect differences in the patterns of exposure (the shorter exposures of 5.7 years versus the longer of 14 years reported by Kunz *et al.*, 1979) as well as an age effect. The age at first exposure as a risk factor is discussed in the model developed to calculate lifetime lung cancer risk described in Section 10.

8.3.3 *Swedish Miners*

Lung cancer mortality among several groups of iron, zinc, and lead miners (Axelson and Sundell, 1978; Jorgensen, 1973; Renard, 1974; Snihs, 1975; and Radford, 1981) has been reported. Radon measurements were made in all Swedish mines in 1969 and 1970 and periodically since then (Snihs, 1975). He converted these measurements to Working Levels assuming that equilibrium existed, that is, 1 WL is equivalent to 100 pCi/L of radon. Concentrations ranged from 0.01 to 200 WL with an average of between 0.5 and 1 WL. Estimates for earlier years were made on the basis of these measurements and on a consideration of changes in ventilation.

Lead and zinc miners of Zinkgruven have been studied by a case-referrent (case control) technique (Axelson and Sundell, 1978; Axelson, 1979). Twenty-nine lung cancer deaths among males (20 in underground miners, NAS, 1980) occurring in the parish from 1956 through 1976 compared with 2.3 expected were used along with 174 other deaths selected close in time to the lung cancer deaths. In the control group, 19 were exposed to radon daughters in underground mines. Since the mines have been in operation since the turn of the century, follow-up in some cases approached 60 years, but lung cancer cases were collected only for the last 20 years of this period. Seven of the lung cancer cases occurred among men who were 70 years of age or older, and only one of them had smoked cigarettes. The mean time worked for lung cancer cases was 32.5 years, and the mean time from cessation of mining to death from lung cancer was 4.1 years. Among the controls who had underground mining experience, 7 smoked out of the 9 who were 70 years or older. Of the cases who died under 70 years of age, 12 had known smoking habits and 3 were reported as nonsmokers. Among controls under 70, 1 out of 7 was a nonsmoker. The smokers had a shorter effective exposure time 18 ± 12 years versus 35 ± 14 for nonsmokers. Stratification using a multivariate confounder score reduced the difference in risk between smokers and nonsmokers, but there still seemed to be a higher lifetime risk of lung cancer among nonsmoking than smoking miners. These data suggested to Axelson and Sundell (1978) that smoking might give some protection against the effect of radon daughters possibly due to increased mucus production. The induction-latent period averaged 34 years for smokers and 43.4 years for nonsmokers ($p < .05$). The population sizes of miners and ex-miners from which the lung cancers were drawn were not known precisely but the person years of observation were estimated as 1718 for the 50–69 year age group and 437 for the 70+ year groups (Axelson, 1979). The expected lung cancers were from general population rates but were obtained by a proportional mortality method, so they are not precise. Similarly, exposures are estimates and NAS (1980) has attributed a cumulative average exposure of 270 WLM for a total of 2154 person years at risk yielding an overall annual risk rate of 30.4×10^{-6} per person per year per WLM.

Kiruna iron miners were studied by Jorgensen with a proportional mortality method (Jorgensen, 1973). Most employees were Laplanders. All male deaths which occurred in the Kiruna district between 1950 and 1970 were collected. The mine had been mostly an open pit operation until about 1950 and subterranean mining was stated to have begun in the late 1950's, so only a few men had exposure follow-up which exceeded 10 years. Expected lung cancer rates were calculated in three ways, including age-specific rates derived from all Swedish males. About 4.5 lung cancers were expected and 13 were observed. Eight of the 13 miners with lung cancer had smoked cigarettes (10–15 per day) and 4 had smoked a pipe only. The deaths were from a group of about 4700 underground miners followed from 1950 to 1970. Since underground mining did not commence until the late 1950's and we assume that it requires about seven years from first exposure to observe a lung cancer, only about seven years for expression of risk or 32,900 person years at risk are available. Radon measurements indicated that levels were from 10 to 30 pCi/L. An assumption of exposure since 1956 yields an average cumulative exposure of about 36 WLM and, therefore, an attributable risk rate of 7×10^{-6} per person per year per WLM.

Using a case control approach, Edling (1982) studied the influence of cigarette smoking on the lung cancer rates of iron miners in Grangesberg, Sweden where underground mining commenced in 1910. Radon daughter exposures were measured as 0.3 to 1.0 WL prior to 1970 and below 0.3 WL after a new ventilation system was installed in 1970. The cases were those subjects who died of lung cancer in the parish during 1957 to 1977 and controls were those who died of other causes. Smoking histories were obtained by telephone interview with next of kin. There was a 16-fold increase in lung cancer among miners, the increase occurring among both smokers and nonsmokers. The most heavy-smoking miners tended to die earlier and to have shorter induction-latency periods for lung cancer. The effect of cigarette smoking and radiation were considered to be additive, not synergistic.

Malmberget iron miners are undergoing intensive study (Renard, 1974; Radford, 1981, 1982). Originally an open pit mine, operations at Malmberget have been completely underground since 1930. The study encompasses a population of 1435 iron miners born between 1880 and 1919 and known to be alive on January 1, 1920. Including only lung cancer deaths occurring ten years after onset of mining (one lung cancer occurred less than ten years after beginning mining) a total of 50 lung cancers have been observed. Of the 50 cases, 32 smoked currently or within the previous 5 years and 18 never smoked or had given up smoking 18 or more years previously. The expected numbers of lung cancers among the smokers and nonsmokers were originally reported as 8.41 and 1.67 respectively based upon Swedish national data collected by the Karolinska Institute, accounting for the proportion and amount smoked in the mining population. The absolute risk was calculated as 22.3×10^{-6} and 16.3×10^{-6} per person per year per WLM for smokers and nonsmokers, respectively (Radford, 1981). The average exposure of these miners was estimated as 97 WLM (5 WLM per year for about 20 years). The attributable lifetime lung cancer risk in this group is therefore $(40/1435) \times 100$ or 3 percent.

In a reevaluation of these data (Radford and Renard, 1981), the expected numbers of lung cancers are estimated as 11.14 for smokers and 1.8 for nonsmokers giving an attributable risk rate of 20×10^{-6} and 16×10^{-6} per person per year per WLM for smokers and nonsmokers, respectively.

Approximately 50 percent of Swedish males smoked cigarettes (Statistika, 1965). However, those who used tobacco were relatively light smokers (the majority smoked less than 15 cigarettes per day). When compared to other Swedish occupational groups, miners had the lowest frequency of regular daily smoking (27 percent) and the highest frequency of occasional cigarette usage (39 percent). For this reason, cigarette smoke probably plays a much smaller role in lung cancer among Swedish miners than in other groups included in Table 8.1.

8.3.4 *Canadian Miners*

Uranium mining began in 1954 in two areas in the Province of Ontario, Elliot Lake and Bancroft. It grew rapidly with more than 15 operating mines and more than 11,000 miners during peak employment years. Bancroft ceased production in 1964 and by 1965 employment was 1351 at Elliot Lake (Chovil, 1981). Exploratory radon daughter measurements were made in 1955 and 1956, routine measurements were begun in 1957 and have continued to the present.

Ontario has, for a number of years, maintained a Uranium Nominal Roll. It lists 15,094 persons who worked for one or more months in uranium mines during the 1955–1974 period (Hewitt, 1976, 1979; McCullough *et al.*, 1979). Eighty-one deaths among this group have been certified as due to lung cancer compared with 45 expected. To assess the role of radiation, a 1 percent sample of survivors (159) was randomly selected. This comparison series was taken to be representative of all persons who entered the mines in the 1955–1964 period and survived to the end of 1974. It should be noted that this was a relatively small sample of the workers who had not developed cancer. Work histories were compared between the two groups. For those who died of lung cancer, the mean time from cessation of work to death was 8.7 years. Cases had an average underground exposure of 3.9 years with 74.5 WLM versus 2.1 years and 32.3 WLM for controls. This was considered as evidence that radiation exposure in mines contributed to the excess lung cancer deaths. It also reflects relatively short exposure periods. Risk calculations based on stratification by WLM disclosed an increase in lung cancer risk with increasing dose. The data were given as proportion of miners born in 1933 or earlier who developed lung cancer by 1974. This age cutoff was used because none of the lung cancers occurred in men born later than this date. The median year of starting underground uranium mining among those who developed lung cancer was 1957. The mean follow-up time was about 15 years (McCullough *et al.*, 1979). Those with highest exposures started mining earlier and were older at observation than were those with low WLM exposures. The average age at start of mining was 29 years, but for lung cancer cases was 40 years. The five stratifications by WLM have been combined here into three to decrease random statistical fluctuation of individual points (Table 8.1). The median WLM value of lung cancer cases was used as the mean exposure for the three categories. This is probably somewhat high as the lung cancer cases had higher than average exposures. Normal lung cancer rates for other groups of active workers in the United States are usually between 300 and 500 per million per year and, in addition, the rate for the lowest exposed group of Canadians (1–29 WLM) was 370 per million per year, assuming a 15-year follow-up (Hewitt, 1976). The expected rate was arbitrarily increased slightly for the higher exposed groups on the assumption that they had started mining earlier and were therefore probably somewhat older at observation, so the estimates for expected lung cancers were 400, 500 and 700 per million per year respectively, for the three exposure groups.

The calculated rates were divided by the estimated observation time to obtain rates expressed in person years. Although these data are useful in establishing an effect at low levels of exposure, the follow-up time is so short that it can contribute little to the present analysis. No information was given on cigarette smoking, but tobacco habits are probably similar to those of U.S. miners because of common cultural influences.

Canadian Miners Traced through the Workmen's Compensation Board

Chovil (1981) reported on lung cancer cases identified through records in the Workmen's Compensation Board of Ontario. He was able to identify 154 workers through 1980 with "both a history of exposure to uranium products and a possible diagnosis of lung cancer." Nineteen of these were excluded because of absence of actual underground exposure or lack of an established diagnosis of primary lung cancer.

Expected lung cancer cases, calculated for the whole cohort at risk using the age distribution of the 1 percent sample (Hewitt, 1976), numbered 85 from 1960 to 1979. The 1 percent sample was restricted to miners beginning employment before 1962 and contained 134 miners rather than the usual total of 159 miners. Expected lung cancer cases in each WLM category can be calculated from the data of Chovil (1981), by apportioning the total 85 cases among the WLM categories according to the 1 percent sample size. This may be done since the age distribution in each WLM category is about the same. Lung cancer cases were recorded over the years 1960 to 1980. The cases through

1974 were acknowledged as reasonably complete but lung cancer cases in later years may not yet have been reported to the Workmen's Compensation Board.

For the purpose of estimating absolute risk, the years over which risk is accumulated are assumed to be from 1960 to 1979. This does not allow a long enough latent interval; furthermore, not enough detail for each exposure category is reported to refine the years at risk. Thus, the absolute risk rate is slightly underestimated. On the other hand, the expected number of lung cancer cases is not reported for 1980 and this will slightly overestimate the absolute risk rate.

The figures reported by Chovil are shown in Table 8.2 in order to indicate the distribution of exposure in the modified 1 percent population sample.

A deficit rather than an excess of cases is apparent in the lowest exposure category which is similar to the U.S. experience so far. These deficits may reflect the "healthy worker" effect (see, for example, McMichael, 1976; Fox and Collier, 1976). Statistically, any negative values must be included when calculating a mean risk so that the risk estimate will be unbiased.

8.3.5 *Other Mining Groups*

The early Schneeburg and Joachimsthal lung cancer experience in Central Europe from the mining of ores for silver, cobalt, bismuth, nickel, radium, and arsenic are adequately summarized elsewhere (Hueper, 1942; Lundin *et al.*, 1971). These were the same ore bodies mined by Czechoslovakian workers described in Section 8.3.2. Lung cancer is reported to have caused between 30 and 70 percent of the

TABLE 8.2—*Attributable risk rate of lung cancer due to radon daughter exposure of Canadian miners calculated from data of Chovil (1981)*

WLM Category	Mean Exposure in WLM	Lung Cancers Cases	Expected Cases[c]	One Percent Population Sample (See Text)	Attributable Lung Cancers/ WLM·y·10^6 Persons[a]
0–30	15	38	56	88	−7
31–60	46	23	14	23	4
61–90	76	20	5	8	12
91–120	106	12	4.4	7	5
121–180	150	21	3.2	5	12
181+	—	21	2	3	[b]
Total		135	85	134	

[a] Assuming risk is expressed over 20 years (1960–1979).

[b] Exposure category not suitable for calculation of attributable lung cancers.

[c] Derived assuming expected cases may be apportioned among the WLM categories according to the 1% sample distribution.

deaths among these miners. Estimates of the lung cancer rates were between 9500 and 18,000 per $\times 10^6$ miners per year, with estimates of mean exposure between about 4000 and 6000 WLM. This gives the estimated rate of attributable lung cancer as about 5×10^{-6} per person per year per WLM.

Extensive studies of Newfoundland fluorspar miners have shown excess lung cancer (Royal Commission, 1969; Devilliers and Windish, 1964; Parsons *et al.*, 1964; McCullough *et al.*, 1979; Wright and Couves, 1977; NAS, 1980; Morrison *et al.*, 1981). Surface mining began in 1933 and underground operations did not commence until 1936. There were five of these mines, but three were closed before measurements were made in about 1960. The source of radon in these mines was shown to be ^{222}Rn rich water which was plentiful in the mines. Samples of mine water were found to contain from 4240 to 12,850 pCi ^{222}Rn per liter. In earlier years (1936–1959), little or no ventilation had been used, but it was being used at the time of the measurements. However, average cumulative exposures were estimated for all the workers and lung cancer rates calculated.

There were a total of 2779 miners in both surface and underground operations during the period 1933–1977. It was reported that 630 miners had underground experience of one or more years. Radon daughter levels in the two remaining mines were measured and found to be from 2.5 to 8 WL. Mechanical ventilation was supplied to one of these mines beginning in 1952 and thus, lower radon daughter levels would be expected there. Improved ventilation in the mines was installed promptly after 1959 when excessive radon daughter levels were measured. The average exposure duration for these miners at the elevated radon daughter levels was about 13 years. Since early underground conditions (late 1930's to 1950's) relied on natural ventilation, radon daughter levels were reported to be from 2 to 8 WL. An average cumulative exposure for the entire group of miners is estimated to be 204 WLM (NAS, 1980).

Up to 1964, 29 lung cancer deaths were observed and 71 up to 1971, 65 among underground workers and as of July 1981, 105 lung cancers had been identified (Hollywood, 1981). Underground mining commenced in 1936 and allowing seven years for the development of lung cancer (the first lung cancer was actually observed in 1949) yields 34 years for risk expression through 1971. McCullough *et al.* (1978) report that 57 lung cancers appeared (1 to 2 expected) through 1976 in the 1000 to 3000+ WLM exposure category. This category included 167 miners. A weighted average (persons × WLM) gives an exposure of 1600 WLM for this high exposure group and an attributable risk rate of $(57\text{-}2)/(34)(167)(1600) = 6 \times 10^{-6}$ per person per year per WLM.

The attributable risk rate for the entire population is estimated to be about 18×10^{-6} per person per year per WLM (NAS, 1980).

British underground iron miners are reported to have excess lung cancer (Faulds and Stewart, 1956; Boyd *et al.*, 1970). There were 36 lung cancers vs. 20.6 expected ($P < 0.001$) among hematite miners. The expected rate, however, was based on "other occupations" as stated on death certificates. Up to 30 percent of the control group may have had some iron mining experience. Over-reporting of compensable disease on death certificates (e.g., pneumoconiosis) to the exclusion of other causes of death such as respiratory cancer seemed to result in an understatement of the number of deaths from cancer (Boyd *et al.*, 1970). Measurements of radon daughters ranged from 0.15 to 3.2 WL (Duggan *et al.*, 1970). No estimates of exposure in earlier years were made, so exposure values would be quite tentative even as the lung cancer rates were. This group was, therefore, not used for our analysis.

An excess of lung cancer has also been reported among French iron miners (Roussel, 1964; Editorial, 1970).

A group of underground hard rock, base metal miners in the United States was reported to have an excess lung cancer rate (Wagoner *et al.*, 1963). Measurements of radon daughters in 1958 in these mines varied from 0.1 to 0.2 WL. Concentrations during earlier periods (before the workers died) are unknown, but were undoubtedly much higher as ventilation had been greatly improved in recent years. These data were felt to be too tentative for use in this analysis.

No excess lung cancer is reported for South African gold miners (Oosthuizen *et al.*, 1958; Chatebakis, 1960). It is difficult to evaluate the reports on the experience in South African gold mines as neither adequate epidemiological data nor radiation measurements are available. The South African Bantu miner life span is only about 40+ years and the white South African miner 58, so there would be little chance of cancer appearing in this group.

Excess lung cancer in Cornish tin miners has been reported (Fox *et al.*, 1981). The population consisted of 1333 miners whose occupational description in 1939 included mention of tin mining. Underground miners exhibited a two-fold increase in mortality over that expected. Radon daughter concentrations were not measured in these mines until 1967–1968 and indicated 3.4 and 1.2 average Working Levels in the two mines studied. Cumulative exposures are not available for these miners and so attributable risk rates cannot at present be calculated.

No excess of respiratory cancer was demonstrated at radon daughter levels below 0.02 WL among United States potash miners (Waxweiler *et al.*, 1973), nor among a group of Russian manganese miners who were rarely exposed to radon daughter levels above 0.3 WL (Gabuniya, 1969).

8.3.6 *Confounding Factors Present in Human Underground Mining Studies*

There are many factors that make it difficult to interpret the human epidemiological data with great accuracy. The most serious defect is that cumulative exposure is not known accurately in essentially all of the follow-up studies. Another significant problem is the observation that the attributable risk per year per WLM is a function of age at first exposure and age at observation. If this is so, then the use of person years in intervals over 10 or more years to calculate attributable risk is not strictly correct. A few of the other difficulties are as follows. Some living lung cancer cases are included in the U.S. study period. The rates are, therefore, not fully comparable with other studies where only mortality is considered. The U.S. mining group has a somewhat higher cigarette usage than that for the general population of adult males. In the Canadian group, cigarette consumption is not known, and the Canadian mine follow-up time is quite short. In the Czechoslovakian mine group the person years at risk in each of seven detailed cumulative exposure groups is not reported. In the Zinkgruven Swedish lead and zinc miners, expected lung cancer rates were determined by proportional mortality, so they are not precise.

In order to accurately assess the number of person years at risk, each person in the group must be followed over a specified interval, accounting for removal due to death from lung cancer or other causes. The minimum latent interval, defined here as the shortest time from first exposure to the detection of lung cancer (about seven years) should not be included in the person years at risk. Similarly, cumulative exposure in WLM should not include exposures for the seven years before the cutoff date for follow-up since these exposures do not contribute to the lung cancers observed for the follow-up period. This detail is often not known.

Along with the difficulties encountered in the epidemiological work, the appropriate bronchial alpha dose can only be estimated from crude estimates of exposure and an average dose conversion factor. The underground atmosphere was not well characterized in any case. In the Czechoslovakian follow-up of uranium miners, 120,000 ^{222}Rn measurements were performed but no radon daughter measurements were reported.

In spite of these difficulties, the Committee on the Biological Effects of Ionizing Radiation (BEIR) (NAS, 1972) also calculated an attributable risk based upon earlier studies and estimated that a uniform appearance rate of 3.2 attributable lung cancers per year per million persons per WLM following a fixed latent interval was an acceptable way to express risk. The 1980 BEIR Committee report (NAS, 1980) reviewed the same studies considered in this report and stated that "risk estimates for lung cancer from radon daughters now range from 6 to 47 $\times$ 10^{-6} per person per year per WLM, and that the range reflects in large part the effect of age at exposure or at onset of the cancer."

From Fig. 8.2 it can be seen that the attributable lung cancer rate is variable with total accumulated exposure in WLM, but there appears to be a decrease in risk per unit exposure at high exposures. This is most noticeable in the U.S. data. This trend was also seen in animal exposures to radon daughters and is discussed in Section 9.

The mean and standard error of the mean of all data reported in the last column of Table 8.1 is 12 $\pm$ 2 $\times$ 10^{-6} per person per year per WLM (excluding the 40+ age group in the Czechoslovakian miners). These data include all studies where reported values of excess attributable lung cancer and cumulative exposures exist. This report adopts a rounded value of 10 attributable lung cancers per million persons per year per WLM noting that a more accurate value awaits completion of the underground mining studies that have been initiated.

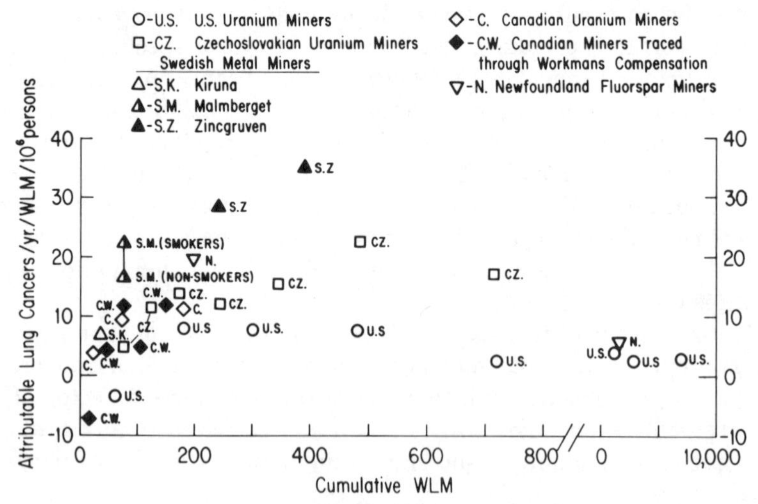

Fig. 8.2. Attributable annual lung cancer risk as a function of cumulative exposure.

8.4 Discussion of Epidemiological Studies

Data on the lung cancer mortality and cumulative radon daughter exposure among selected mining groups are presented in Table 8.1 and Fig. 8.2. The data from the U.S. and Czechoslovakian miners are reasonably similar below about 300 WLM. The Czech data may be better at low cumulative exposure because of a greater number of person years and longer follow-up. The Czech miners first exposed at older age (40+) have a significantly higher risk than those first exposed between the ages of 30 to 39.

In the reported analyses of lung cancer among these mining groups, a number of important factors involved in the relationship between radon daughter exposure and resultant lung cancer in man are becoming apparent. The latent period seems to vary inversely with age at first exposure, with amount of cigarette smoking, and with total exposure and/or exposure rate. That is, the shortest latent periods are found among those men who are elderly at start of mining, who smoke heavily, and who have the most intense exposures. The latent period has a large range of about 7 to 50 years. Mean values are usually considered to be between 20 and 30 years, but one study reported a mean of 43 years among nonsmokers exposed at low levels who had been followed for over 60 years (Axelson, 1960). In none of the studies so far has there been any significant appearance of lung cancer before age 40. This indicates that the concept of a fixed latent interval is somewhat artificial since radon daughter induced lung cancer tends to appear at the ages normally anticipated, although many of the cancers do appear between ages 40 and 55 among miners who smoke.

The incidence of lung cancer among a group of exposed miners is largely dependent on total dose, but there is probably a dose-rate factor involved as noted. Incidence appears to vary directly with age at first exposure, with number of cigarettes smoked, with the age of individuals at time of observation, and with the follow-up time (length of observation period). There may be a paradoxical effect of cigarette smoking. During periods of relatively short follow-up (15–25 years) cigarette smoking is associated with a markedly increased incidence of lung cancer in miners. During periods of follow-up that are 30 to 60 years after initial exposure, lung cancer incidence is reported to be either somewhat greater among nonsmokers than smokers (Axelson, 1980), or about the same (Radford, 1981). A recent report (Gottlieb and Husen, 1982) noted 16 cases of lung cancer among a small group of Navajo uranium miners. Only two of them smoked cigarettes regularly.

Data in Fig. 8.2 are not suitable to determine whether a threshold exposure for lung cancer induction exists. In order to estimate lung cancer risk in exposed populations it is in keeping with present day views of radiation biology and radiation protection to assume that radiation induced-cancer is a stochastic process correlated with dose and without a threshold.

As indicated above, many of the "expected" values for lung cancers in Table 8.1 could not be calculated precisely. For all but the lowest exposure levels such imprecision is probably of little importance as the observed number of lung cancers is markedly greater than the expected. However, this lack of precision does decrease confidence in the data below 200 WLM. This is the region where follow-up has generally been inadequate. In addition, the "healthy worker effect" should be considered in assessing the expected lung cancer rates. That is, among a group of workers (subjected to both medical and self screening) one expects to find lower mortality rates from all causes than among the general population, if their health has not been adversely affected by the work (McMichael, 1976; Fox and Collier, 1976). This factor may be of little importance among those groups with longest follow-up, but is likely to have considerable influence in those groups with short follow-up.

None of the studies, so far, has produced data which show a statistically significant excess of lung cancer in the lowest cumulative exposure category (<60 WLM). Therefore, in estimating the effect of radon daughter exposure at environmental levels, normally less than about 20 WLM per lifetime, the attributable risk at high exposures must be extrapolated to the low exposure region.

8.5 Summary of Epidemiological Findings

A review of epidemiological data on underground miners indicates that an excess lung cancer mortality exists above cumulative exposures of about 100 WLM. Radon daughter exposures appear to be more efficient in inducing lung cancer when cumulative exposures are below 1000 WLM.

All of the reported studies are deficient in several ways. These deficiencies are discussed in sections 8.3.6 and 8.4. Usually there are insufficient data on individual radon daughter exposures and there is not enough detail to determine whether person years at risk are properly accounted for. For this reason, only crude estimates of attributable risk are possible and these range from 1.5 to 45 $\times 10^{-6}$ with a rounded average value of 10×10^{-6} per person per year per WLM. This average value is adopted in this report.

A number of factors have been identified which influence the length of the induction-latent period and the attributable risk of lung cancer appears sensitive to age at first exposure. Lifetime risk, rather than incidence rate, is a more useful way to express total risk since it accounts for the variables involved. Lifetime risk is calculated in Section 10 using the average value of attributable risk adopted here of 10×10^{-6} per person per year per WLM and differing ages at first exposure and exposure duration.

9. Studies of Radon and Radon Daughter Inhalation Using Experimental Animals

9.1 Background

Investigations into the potentially harmful effects of radon began shortly after the discovery of radium itself, with papers appearing in the French (Bouchard *et al.*, 1904) and German (London, 1904) literatures at the beginning of the 20th century. During the ensuing 50 years, sporadic attempts were made to determine the possible mechanisms of lung pathology caused by radon, but these studies usually consisted of brief exposures using very few animals, and were hampered by lack of recognition of the dominant role of the short-lived radon daughters in the production of radiological dose.

Although the existence of lethal pulmonary disease among metal miners of Schneeburg, in Saxony, had been known for centuries (Dibner on Aricola, 1597), the work of Jansen and Schultzer in 1926 is described by Morken (1955a) as the first experimental animal study undertaken in response to the high incidence of lung cancer among uranium miners. Jansen and Schultzer (1926) exposed rats continuously to an atmosphere containing radon at 0.07 μCi/L under static conditions. No deaths in immature rats or adults occurred in 1 or 2 weeks, but 1 of 2 immature rats died 3 weeks following 3 weeks of exposure. One rat died at 4½ weeks of exposure and 2 adult rats died at 4 weeks and at 6 weeks.

Included in Morken's (1955a) survey of studies conducted in the 1930s and 1940s are those of Read and Mottram (1939), of Jackson (1940), and of Rajewsky *et al.* (1942a, 1942b). Read and Mottram exposed small numbers of mice (six per group) continuously to radon concentrations of 1 μCi/L, 0.1 μCi/L and 0.5 μCi/L for 5 days/week.

At the 1 μCi/L concentration, all animals died between 19 and 56 days, but at 0.1 μCi/L all mice tested survived without adverse effect for 161 days. At 0.5 μCi/L, the first animal died on the 35th day. The affected mice showed weight loss after two weeks and some showed hypertrophied spleens. In these and the preceding studies, radon daughter products were present in unknown concentrations, perhaps reaching equilibrium with radon.

In 1940, Jackson reported the results of studies conducted with mice that had been exposed under conditions of radon recirculation at high flow that may have removed most of the attendant radon daughters (Morken, 1955a). A total of 60 mice of strains C57B, C3H, and A were exposed to 0.3 μCi/L for 6 weeks, resulting in 7 deaths. However, spontaneous illnesses may have accounted for 6 of these 7 deaths, and the author expressed concern over the high incidence of spontaneous pulmonary tumors characteristic of A strain mice.

In 1942 and 1943 a series of experiments with mice was reported by Rajewsky and coworkers (1942a, 1942b, 1943). Experiments conducted for 210 days showed no effect other than slight weight loss at 1.35 μCi radon/L (vs. 3 dead from spontaneous illness in controls), but 100 percent deaths at 23 to 71 days of exposure to 20 μCi/L. These deaths occurred at concentrations 20 times higher than those used by Read and Mottram, leading Rajewsky *et al.* (1942a) to suggest much greater sensitivity in immature mice used by the former. Rajewsky *et al.* (1943) presented data showing that at 16 μCi radon/L, immature mice had a mean lifespan of 17 days vs. 99 days for adult animals; some hyperplasia of bronchial epithelium was noted. In an effort to produce cancer, Rajewsky *et al.* (1942b) selected 1.16 μCi radon/L as a level that might cause pulmonary effects without shortening lifespan. Twelve mice were exposed continuously for 161 to 453 days. Mean lifespan was found to be 286 days; three-fourths of controls lived beyond this time. The authors suggested a possible increased incidence of lung cancer in these mice, but their findings were weakened by the occurrence of spontaneous illness and bronchial epithelial thickening in control animals.

Experiments performed prior to the early 1950s generally involved exposure periods of only a few days to a few months, and usually were based on radon concentrations, with little or no consideration of possible lung irradiation by accompanying radon daughters. However, during 1952–1957 there emerged a growing concern over the possibility of increased lung cancer among uranium miners of the Colorado Plateau (Holaday *et al.*, 1952; Bale and Shapiro, 1956), and the importance of accurately determining the levels of radon daughter radionuclides in mine air was pointed out by Holaday (1952), Harley (1952), Shapiro and Bale (1953), and Harris (1954). Several investigators at the University of Rochester began to focus attention on the biological effects (Morken, 1955b; Morken and Scott, 1966) and physical behavior (Bale, 1951) of radon daughters per se, and the contribution of these decay products to the overall radiation insult to various portions of the respiratory tract (Holaday *et al.*, 1957). Shapiro and Bale (1953) exposed rats and dogs to several levels of radon, in the presence and absence of radon daughters attached to aerosol particles in "dusty air." In 1956, they compared levels of individual radon daughter radionuclides measured in the trachea with calculated values based upon models of the human respiratory tract, and showed that the radon daughters attached to carrier dust particles were a primary factor in the resulting alpha radiation dose to the airway epithelium at various levels in the respiratory tract (Shapiro, 1956). They also postulated that this dose was primarily due (>95 percent) to radon daughters RaA (^{218}Po) and RaC′ (^{214}Po), rather than to the parent radon.

Cohn *et al.* (1953) reported the levels of radioactivity found in three regions of the respiratory tract of Sprague-Dawley female rats exposed to either radon alone or radon in equilibrium with its decay products. Groups of six animals received three-hour head-only exposures, and resulting gamma counts in the nasal passages, trachea plus major bronchi, and remainder of lungs were compared. Findings of 125 times greater gamma activities (RaB and RaC) in the respiratory tracts of animals that inhaled radon plus its decay products compared to animals that inhaled radon alone, together with greatly increased alpha radioactivity from the radon daughter RaC′, led to the conclusion that uranium miners should wear masks to filter out radon daughters, and that "inhalation studies which consider the radiation from radon gas alone do not provide a realistic approach to the problem insofar as hazards to humans are concerned." These authors also were among the first to show the importance of absorbed dose (at a specific deposition site) rather than exposure, and to suggest that a much higher radiation exposure per unit area occurred in the bronchi compared with other lung tissue.

Following the earlier studies, three comprehensive investigations with experimental animals have been conducted at the University of Rochester (Section 9.3), the Centre d'Etudes Nucleaires (CEN) in France (Section 9.4). and Battelle Pacific Northwest Laboratory (Section 9.5). These studies are described in detail in the sections noted. Also, the detailed dosimetric study performed at Harvard (Section 9.2) for the Syrian golden hamster is described. It should be noted that the hamster respiratory tract is somewhat similar to that of the rat.

9.2 Absorbed Dose From Inhaled Radon Daughters in the Syrian Golden Hamster

Desrosiers *et al.* (1978) obtained morphometric (silastic cast) and histologic descriptions for the Syrian golden hamster respiratory tract. This information was used in calculations of mucus velocity and particulate deposition in the trachea. Nine generations of conductive airways plus the parenchymal region of the lung were considered during the postulated inhalation of 1 liter of a reference radon atmosphere, and the number of RaA and RaC′ disintegrations were calculated for each level of the hamster lung.

9.2.1 *Methods*

Measured airways were classified into groups: trachea as Group 1, bronchi were found in Groups 2 through 7, bronchioles in Groups 4 through 9, and terminal bronchioles were found in Groups 4 through 10. Each terminal bronchiole was followed by an average of 2.4 respiratory bronchioles and these subsequently branched into six alveolar ducts and eight alveolar sacs.

Calculations of the mucus velocity were based upon the technique described by Altshuler *et al.* (1964). A mucus thickness of 12 micrometers was selected on the basis of electron micrographs of hamster respiratory epithelium. The report does not evaluate the possibility of alteration of mucus layers during fixing for electron microscopy.

The authors assumed 10 to 30 percent unattached RaA as suggested by Haque and Collinson (1967) and by Raabe (1977). This corresponds to conditions at the University of Rochester in which 20 percent unattached RaA was used (see Section 9.3), and to studies at Battelle Pacific Northwest Laboratory in those chambers without added uranium ore dust or diesel engine exhaust particles (see Section 9.5).

The retention efficiencies calculated by this model correspond roughly to measurements of radon daughters in rats made by Shapiro (1956), as described above. The authors recognized that nasopharyngeal deposition of either unattached radon daughters or daughters attached to very fine particles is extremely difficult to estimate and pointed out that fine details of the anatomy of the region would very likely increase the nasal deposition. There is no further development of the consequences of such nasal depositions. However, this effect may be important because squamous carcinoma was produced in nasal passages of dogs, hamsters and rats in studies at Battelle Pacific Northwest Laboratory, though not reported in humans.

9.2.2 *Results*

Desrosiers *et al.* (1978) calculated the airway dose for three reference mine atmospheres. For an atmosphere containing radon daughters in equilibrium with radon with 4 percent unattached RaA and a particle size of 0.3 μm, the doses to regions of the lung were calculated as follows; 0.2 rad/WLM to the trachea, 0.03 to 0.05 rad/WLM to the bronchi and bronchioles, 0.4 to 0.5 rad/WLM to the finest bronchioles, 0.1 rad/WLM to the terminal bronchioles and about 0.07 rad/WLM as an average for tissues in the respiratory region. This range of dose rates in hamsters is much greater than is calculated for humans and, for this reference atmosphere, the overall magnitude of dose rates to several regions is much less in hamsters than would be the case in humans.

Altshuler *et al.* (1964) described a dose range of 1 to 2 rad/WLM in humans, depending on whether nose breathing or mouth breathing was involved. The model described by Desrosiers *et al.*, utilizing the Altshuler reference atmosphere, yields 0.1 rad/WLM to the hamster basal cell of the 4th airway group and 1.4 rad/WLM for basal cells in the 8th group. The reference atmosphere of Altshuler *et al.* (1964) contained about 10 percent unattached RaA and the majority of activity was attached to very small particles, i.e., 0.03 μm diameter. The remaining atoms were distributed over particles of 0.2, 0.6, 1.2 and 2 micrometers diameter. The dose to the terminal bronchioles relative to generation 4 probably explains the production of squamous carcinoma in the peripheral lung in the Battelle Pacific Northwest studies in hamsters.

The third reference atmosphere was based upon that used by Haque and Collinson (1967) to simulate a "clean room" radon daughter exposure situation. In this case, radon daughters were assumed to be unattached or absorbed on particles ranging from 0.1 to 0.4 micrometers in diameter. Estimates of absorbed doses to the subsegmental bronchi in humans indicate, with Desrosiers' corrections, a dose of 4 rad/WLM. Applying this reference atmosphere to the hamster lung model yields a dose of 0.2 rad/WLM to the alveolar regions of the lung and a dose of 0.8 rad/WLM to the basal cells in generation 8.

Desrosiers *et al.* (1978) show that all reference atmospheres produce radiation doses in the hamster that are lower in the upper airways than those calculated in humans, and that the distribution of the absorbed dose in the case of hamster "spares" those regions which are most affected in man.

9.2.3 *Discussion*

Desrosiers *et al.* (1978) conclude from these calculations that when exposure to the 4th airway group of basal cells is averaged over these three reference atmospheres, the resultant dose is approximately 0.08 rad/WLM. They therefore suggest that larger amounts of inhaled radioactivity will be required to produce cancer in hamster lungs than are implicated in human central airway carcinoma. In addition, all three estimates of the human exposure place the greatest radioactivity levels in the segmental bronchi, whereas in hamsters the most peripheral areas (including terminal bronchioles and alveolar regions) and the trachea receive a dose per unit exposure that is comparable with that received by humans.

The mean dose to all basal cells was calculated to be 0.1 to 0.3 rad/WLM considering results from all three reference atmospheres. The mean dose received by Clara cells and terminal bronchioles also ranges from 0.1 to 0.3 rad/WLM.

In the case of the uranium mine atmosphere postulated by Altshuler, the dose of 0.1 rad/WLM to the 4th group in hamster lungs is about an order of magnitude less than absorbed dose factors calculated for the upper airways of humans (generation 4). This may be a partial explanation of the absence of bronchial carcinoma in hamsters in studies conducted by Battelle Northwest Pacific Laboratory, as described in Section 9.5. No discussion is given as to whether the absorbed dose model in hamsters shows better correspondence to nasal vs. mouth breathing in humans.

9.3 Radon Daughter Inhalation Studies Using Experimental Animals at the University of Rochester

Studies were performed by Morken (Morken, 1955b; Morken and Scott, 1966; Morken, 1973a, 1973b) to determine possible relationships between exposure of experimental animals (mice, rats, dogs) to radon and radon decay products and the ultimate development of lung tumors. These experiments involved about 2000 mice, 100 rats, and 80 dogs, plus controls.[4]

9.3.1 *Exposure System*

Unfiltered room air (25 degrees C and 50 percent R.H.) entered the top of a Rochester exposure chamber (2 m^3) and was exhausted from the bottom. The unfiltered air presented the animals with a natural aerosol at a dust load of about 0.1 milligram per cubic meter. Radon was admitted to the top of the chamber to be mixed with the incoming air and dust. The radon was generated at the rate of 22 microcuries per minute from a solution of radium-226 in hydrochloric acid and was swept from the solution into the chamber by an air stream.

The radon concentrations were about 0.5 microcurie/liter for the mouse and dog studies and about 1.0 microcurie/liter for the rat study. The aerosol itself contained a variety of dusts of generally less than 1 micrometer diameter, plus oil and water droplets. The median aerodynamic diameters of particulate activity were consistently close to 0.2 μm, and the fraction of unattached RaA activity was about 20 percent (Mercer, 1970).

9.3.2 *Experiments Using Mice*

CAF strain mice were 20 weeks old at the beginning of exposure. These mice were exposed for three 50-hour periods each week for intervals which ranged from 8 weeks to life. The chamber air flow of 45 liters per minute resulted in steady-state concentrations for radon, RaA, RaB, and RaC of 0.50, 0.45, 0.23, and 0.13 microcuries per liter, respectively. This represented a daughter concentration of 2000 WL and an exposure rate of 600 WLM per week.

Measurements on mice exposed 24 hours to attain steady state showed the following average dose rates from alpha radioactivity:

1) whole body, 5 rad/week,
2) kidney, 18 rad/week,
3) liver, 2 rad/week,
4) G.I. tract, stomach and contents, 60 rad/week,

[4] Histopathology examinations were performed on mice by Dr. J.K. Scott, on rats by Dr. C. Yuille, and on dogs by Dr. G. Casarett and Dr. C. Yuille.

5) lung-trachea-bronchi, 280 rad/week.

Dose to bronchial tissue may have been 5 to 10 times that to the entire lung, or as much as 2800 rad/week.

In the first experiment, the mice were exposed continually throughout life. The median exposure lifespan was 35 weeks, a 50 percent shortening of normal lifespan. The trachea displayed mucosal lesions that were hyperplastic, metaplastic, and destructive; often all of these lesions were observed in the same animal. Squamous metaplasia was manifested as a thickened mucosa and sometimes included keratinization. Lesions observed in the major bronchi were similar to those in the trachea. The observed hyperplasia appeared the same, although squamous metaplasia was not as marked in this tissue.

Intra-alveolar edema was present in varying degrees in all animals. Foci of alveolar phagocytes were seen in all animals; some phagocytes contained a brown pigment. A few small foci of adenomatoid proliferation of bronchi accompanied with fibrosis were seen in a few animals. The lesions of the bronchial mucosa were summarized as destructive, hyperplastic, and metaplastic.

No carcinoma was found but many of the changes were considered precancerous. The exposures for these mice amounted to 72,000 WLM and the average lung dose was about 11,000 rad.

In the second experiment, groups of mice were exposed for 15, 25, and 35 weeks resulting in exposures of 27,000, 45,000, and 63,000 WLM, respectively, producing lung doses from 4200 to 9800 rad. The 35-week exposure confirmed the median lethal dose of the previous experiment, but life shortening was not found in the groups exposed for 15 and 25 weeks.

In the third experiment, groups of mice were exposed for 10 weeks, 15 weeks, 20 weeks, and 25 weeks. Mice were sacrificed after the end of exposure and then at 10-week intervals at 60 weeks of age until 110 weeks. The exposures ranged from 18,000 to 45,000 WLM, producing lung doses from 2800 to 7000 rad.

The pathological effects found in the 2nd and 3rd experiments were not marked. Lesions in the trachea and bronchi were similar to, but not as extensive as those observed in the 1st experiment. The extent of the lesions seen immediately after exposure appeared to be proportional to the exposure duration (cumulated exposure dose). Repair was rapid and by eight weeks postexposure the tissues appeared normal. Changes in the large bronchi were less marked than in the trachea. The terminal bronchioles showed atypical changes in the cuboidal lining epithelium which tended to flatten or disappear with increased post-exposure time.

The tumors found were variants of those which occur spontaneously in this strain of mouse (adenomas and foci of adenomatosis) and were seen in some of the older control mice. No tumors were found after 25 weeks of exposure in any animal sacrificed between 25 and 100 weeks postexposure. Most of the adenomas showed qualitative changes suggestive of a more malignant behavior; i.e., greater cellular anaplasia, a high mitotic rate, and invasion of adjacent lung tissue.

Small microscopic sized neoplastic lesions were found in mice sacrificed up to eight weeks after the 35-week exposure and most of these were of a type not seen after the lower doses. These consisted of ill-defined areas containing many small, often intra-alveolar, nests of bizarre cells. Occasionally these were found in several lobes; some cells appeared to be squamous and associated with keratin. Other tumors in this group had a glandular pattern with local malignant features. These tumorlets did not appear in other exposure groups of mice but have since been observed in exposed rats (see below, Section 9.3.4).

9.3.3. *Experiments Using dogs*

Individual exposure periods for dogs were 20 hours/day, 5 days/week. In some instances, exposures were made on weekends to obtain daily consecutive treatments.

The first dog experiment was performed with one exposed and one control dog. The radon concentration was about 0.26 μCi/liter with an air flow of 45 Lpm. Total exposure duration was 45 days (9 weeks). The exposure level was 220 WL; 900 hours of exposure (cumulated) resulted in a total of 1200 WLM. Estimated dose was 140 rad to whole lung and 4000 rad to tracheal bifurcations.

The trachea showed epithelial squamous metaplasia without keratinization. No changes which could be attributed to radon inhalation were seen in the lung except for some areas of minimal epithelial thickening.

Because only a slight effect appeared in the first dog experiment, a second experiment was performed as shown:

Group	No. of Dogs	Exposure Schedule	WL	WLM	Tracheo-Bronchial Dose	Sacrifice At
B	2	5d/wk for 2 wks	1660	1950	6500 rad	6 mos.
B	2	5d/wk for 2 wks	1660	1950	6500 rad	12 mos.
C	1	1d/wk for 10 wks	1660	1950	6500 rad	6 mos.
C	1	2d/wk for 10 wks	1660	3900	13,000 rad	6 mos.

Control dogs were not used

The histopathologic changes observed were subtle and occurred in small microscopic foci that were usually widely separated. The changes most frequently observed were foci of subacute or chronic inflammation in small bronchioles in alveolar walls. These foci showed fibroblastic activity and fibrosis and collections of lymphocytes but very few polymorphonuclear leukocytes. The inflammation was more frequently acute at 6 months, but chronic at 12 months postexposure. Foci were much less frequent and of lesser degree in C group than in B group, although occasional foci were more prominent and larger in C group. Changes in tracheobronchial and bronchiolar epithelium were relatively infrequent and consisted chiefly of foci of slight epithelial hyperplasia.

There was no appreciable difference between the two dogs of C group which received different cumulative exposures. It seemed possible that exposure to radon did cause some of the subtle inflammatory changes, but adequate evaluation of such subtle changes required a study with control dogs.

The National Institute for Environmental Health Sciences supported a pilot experiment with 51 dogs in a pathology study over a three-year period. Nine other dogs were used in a dosimetry-only experiment. Dogs were exposed for 1, 2, 4, 8, 15, and 50 days to provide cumulative exposure, at the rate of 200 WLM per day, from 200 to 10,000 WLM. Forty dogs were exposed to radon daughters, four chamber-control dogs were exposed for 50 days under similar conditions but without radon daughters, and 7 age-control dogs received no chamber exposure. Dogs were sacrificed at 0, 1, 2, and 3 years postexposure.

At the prescribed time, the dog was sacrificed and the lung fixed *in situ*. A prescribed protocol provided samples from lung, respiratory tract, cervical and submandibular lymph nodes, salivary gland, tonsils, larynx (soft as well as cartilaginous), mucous membrane of the pharynx, pulmonary lymph nodes, and nasal conchae. Gross examination was made of all major organs. No visible evidence of tumors or other lesions was seen in any organ or tissue.

Alveolar doses ranged from 34 to 1700 rad; tracheal and bronchial doses ranged from 1000 to 50,000 rad. No dogs died during or following exposure until sacrificed in the histopathology study. All dogs remained in good health during the five years of this study.

No cancers or tumors were found. Pathologic changes were found in lung sections.

a. The microscopic lesions observed, for all doses, were subtle, variable, very small spatially, widely separated, and involved only a very small fraction of the lung.

b. There were no significant differences between age-control and chamber-control dogs with respect to nonparasitic lesions.

c. Dose-effect relationships:

(1) Immediately after the end of exposure: no relationship and no significant differences from control groups were found for groups exposed from 200 to 10,000 WLM.

(2) An increase of small foci of chronic inflammation was found as the exposure progressed from 200 to 10,000 WLM at 1 and 2 years after exposure. This relationship disappeared at the lower exposure by the third year and was equivocal at the larger doses.

d. The lesions at all doses appeared as small patches of thickened alveolar walls with some metaplasia of alveolar cells and some hyperplasia of bronchial epithelium.

e. The author felt that the extent to which lesions either repair or progress in severity may determine their potential importance in pulmonary carcinogenesis. The lack of lesions in the upper bronchial region allowed no conclusion as to the probability of bronchial carcinogenesis.

An estimate of the radiological dose to lung was obtained from experiments with nine dogs. Each dog was exposed for about 20 hours (to ensure the steady state situation), then killed immediately. The lungs were removed within a few minutes and sectioned into trachea, major bifurcation, and lung proper. All sections were weighed and the RaB and RaC activities were determined over a period of two hours. The counting data, analyzed by decay curve methods, were compared with similar data from dust samples collected before the dog was removed. A summary of results from these dogs is:

a. Average dose rate to whole lung amounted to about 0.17 rad/WLM. Individual lung doses ranged from 0.08 to 0.79 rad/WLM.

b. Clearance rate estimates ranged from half-times of 20 to 198 minutes for whole lung, 17 to 356 minutes for lung halves, and 11 to 231 minutes for lung lobes. The average clearance half-time for whole lung was 35 minutes.

c. From the measured beta activities, the calculated alpha dose rates in the trachea ranged from 0.28 to 12.5 rad/WLM (4.7 ± 4.4) and those for the bifurcation ranged from 0.36 to 20.7 (5.0 ± 6.2). These dose factors for the epithelium (not basal cells) were estimated from the radioactivity measured, the dimensions of the sections, and an assumed 50-micrometer layer for alpha particle absorption.

Clearance rates in these sections were so large that no accumulation of decay products occurred above air concentrations, and the relative activities of RaA, RaB, and RaC were assumed equal to that in the air breathed. With this assumption, the author considered the alpha dose rate from RaA to be 3 to 4 times that from RaC′. This is an oversimplification and the dose from RaA is unlikely to be higher than that from RaC′.

The average rad/WLM to the bronchial tree, if similar to that in the trachea, was about 30 times that for the alveolar region. This would be about 5 rad/WLM, which is 10 times higher than the estimate for humans.

9.3.4. *Experiments Using Rats*

For this study, 96 male rats each of standard (STD) and specific-pathogen-free (SPF) Sprague-Dawley strains were used. Forty-eight rats of each type received exposure to radon daughters for 5 hours each day, 5 days each week, to a total of 600 hours (24 weeks). Twenty-four rats of each kind received sham exposures for 600 hours, and 24 rats of each kind received no exposure.

The air flow was adjusted to produce a steady-state radon concentration of 1 μCi/L for 5 hours. For short exposures the cumulated exposure was related linearly with time to the steady-state values of radon and daughters. The chamber operated at a steady-state concentration of 7300 WL and the daily exposure was 214 WLM. In 600 hours the cumulative exposure was 25,800 WLM. If the factor of 0.17 rad/WLM found for the dog (and for the mouse in previous experiments) is applied, the average dose to lung is about 4300 rad. If the factor of 30 (ratio of trachea to lung dose) applies, the trachea may have received a dose of 130,000 rad.

The pathology findings were not striking. For the STD rats, all groups including controls displayed varying degrees of pneumonia, bronchiectasis, bronchitis, congestion and, infrequently, slight squamous metaplasia in the trachea. Additional effects seen in the irradiated STD rats were an increase in the amounts of metaplasia, denudation of the respiratory tract, and a slight fibrosis. Also found were an adenoma in one rat and a squamous tumorlet in another. None of these effects seemed related to dose.

For the SPF rats, the observed lesions in the lungs of all groups were similar to those of the STD rats, including some metaplasia. No squamous metaplasia was seen in the respiratory tract of any irradiated animal; a slight fibrosis was seen in one. Tumorlets, similar to those seen in the earlier experiment with mice, appeared in irradiated SPF rats, 2 occurred at 7 weeks of exposure and 1 at 12 weeks. None of these effects appeared to be dose-related or to progress with time. Post-exposure lungs showed a mild pneumonia. Late post-exposure findings included one undifferentiated carcinoma, one adenoma, and one well-differentiated, keratinizing, squamous carcinoma which occupied nearly all of one lobe of the lung. Dose-effect relationships here are questionable.

9.3.5. *Discussion*

These experiments were done to determine effects of alpha irradiation on lung tissue. The paucity of early and late effects coupled with the large radiation doses delivered to these tissues were results which Morken considered important in several aspects. While injury could be done to the bronchial tree, the lesions did not lead to bronchial tumors or cancer even with continued irradiation and, after irradiation ceased, these lesions were quickly repaired. Late effects did not appear in the bronchial tissue. In the alveolar region, the injury was not immediately evident, but appeared later in life, whether or not irradiation continued. Major, perhaps permanent, changes occurred in the region of the respiratory bronchioles.

The essentially negative character of the results seen suggested that alpha irradiation was particularly inefficient in producing radiation-specific tumors in the lung or respiratory tract. Since the late effects seen were similar to those reported to follow radon-only (no decay products or dust) inhalation at a lung dose of 300 rad (Scott, 1955), a large uncertainty exists concerning the role played by the massive lung doses provided by the inhaled decay products of radon.

The relative failure of these doses to produce even microscopically visible evidence of radiation-specific tumors in the lung or to produce permanent alterations in the respiratory tract which could lead to bronchial carcinoma is striking. The only apparent late and permanent changes occurred in the alveolar regions of the lung for a wide range of doses and for observation times to three years in the dog and one and two years in the rat and mouse. The alveolar changes might eventually have produced lung cancer of peripheral origin but a combination of life span shortening and sacrifice of the animals before tumors could develop probably precluded finding these lesions.

The authors concluded that it appeared to be extraordinarily difficult to produce even small long-lasting effects with alpha irradiation by radon and its daughters.

9.4 Studies of the Biological Effects of Inhaled Radon and Radon Daughters at the Centre d'Etudes Nucleaires

These studies (Chameaud *et al.*, 1976, 1978) were designed to determine whether radon and its daughters could induce tumors in rats, and to provide experimental evidence supporting the epidemiological data that exposure to radon and its decay products contributes to the etiology of lung cancer in uranium miners (Chameaud *et al.*, 1976).

9.4.1 *Materials and Methods*

Prior to 1972, radon was obtained from very rich ore containing 25 percent uranium, ground to fine powder, and spread out on trays in a steel container. Air passed through the trays and carried the radon by a closed circuit to a room containing animal inhalation chambers. Under these conditions, the equilibrium of the daughters with radon was about 30 percent and the radon concentration was 0.75 μCi/liter. Using a filter and an electrostatic purifying apparatus, radon equilibrium with its daughters could be reduced to about 1 percent. The inhalation chambers contained a maximum of 100 animals that could be exposed for about five hours at a time.

A new whole body exposure apparatus has been used since early 1973. In this apparatus, radon was emanated from two stainless steel tanks placed underground, having a total volume of 10 m^3. These contained 57 barrels of radium-rich lead sulfate, each containing 2 curies of radium. Pipes connected the tanks to an equilibration container of 1 m^3, which, in turn, was connected to two metal inhalation chambers of 10 m^3 each. In the studies described here, the required quantity of radon for each exposure was drawn into the 1 m^3 container, where a centrifugal ventilator then distributed the radon into either or both inhalation chambers. At the end of each inhalation exposure period, a fan cleared the chambers of radon. Each of these chambers could contain 300 rats for hours. By adding oxygen, the exposure could last as long as 16 hours.

This apparatus permitted daily exposures with a maximum radon concentration of 1.25 μCi/L and 100 percent equilibrium with radon daughters. For the lower doses with filtered radon, the fractional equilibrium levels of radon decay products were about: $R_aA = 0.042$, $R_aB = 0.0066$, $R_aC = 0.004$, for 0.75 μCi/L of Rn. The total exposure varied from 50 to 600 hours, delivered over a period of 1–10 months.

In two experiments, rats were exposed to either stable cerium hydroxide or to uranium ore dust, as well as to radon and radon daughters. In one experiment, animals were exposed to aerosols of stable cerium hydroxide in a single period before exposures to radon and radon daughters: 0.5 to 1 mg of cerium hydroxide was deposited in the lungs. In the other experiment, animals were exposed to uranium ore dust (15 percent uranium, 130 mg/m^3 for a total of 51 five-hour daily periods), and to radon and radon daughters on alternate days. Male Sprague-Dawley SPF rats were used in these studies. The lungs of the rats were observed for lesions after spontaneous death; or after sacrifice because their health seemed seriously impaired or because they had suddenly lost weight; or, according to some protocol, at various periods after the beginning of exposure (from the 6th to the 24th month).

9.4.2 *Results*

Approximately 800 rats were used in these studies: about 400 lungs were examined microscopically, as well as macroscopically. A total of 250 benign and malignant tumors were found.

The pathological changes caused by chronic inhalation of high levels of radon and radon daughters included large areas of diffuse interstitial pneumonia with formation of hyaline membranes. There was often severe fibrosis of interalveolar septae surrounding the capillaries. Death generally occurred between a few weeks and a few months.

When the cumulative dose was lowered, the animals lived longer and lesions appeared that were classified into four types of pathology. All four types appeared to be related to time after exposure:

1. Two varieties of metaplasia were found. Bronchiolar metaplasia consisted of large columnar cells, with basal nuclei, light-colored protoplasms, often ciliated, and was generally found at the bronchiolo-alveolar junction and in the neighboring alveoli. Alveolar metaplasia of cuboidal cells, with darker protoplasms, generally appeared in the peripheral lungs.
2. Adenomatous lesions were also found. These formed areas of variable size, never very dense, in which the alveolar septum, covered with one or several layers of cells, was easily recognizable.
3. Adenomas found were round-shaped tumors made of cells, often clustered together, in which the seriously-damaged alveolar septum could be seen only after special staining.
4. Malignant tumors of several different types were found. These included epidermoid carcinomas that were not always clearly differentiated and were often keratinized or necrosed and occasionally extended into the mediastinal cavity; bronchiolar ade-

nocarcinomas which were often mucus-producing, invaded a whole lung lobe or even several lobes, contained numerous cellular anomalies and were characterized by a high number of mitoses; remained limited to the lungs for a long time and were seldom metastatic; and bronchiolo-alveolar adenocarcinomas which exhibited very few mitoses.

There was a whole range of intermediary lesions difficult to classify in one or another of the above categories. Some adenomatous lesions were clearly different from carcinoma, but some adenomas showed malignant-looking characteristics. There were tumors with combinations of epidermoid carcinoma and bronchiolar adenocarcinoma. Epidermoid and bronchiolo-alveolar carcinomas were found in the same animal and occasionally in the same lung. A large number of malignant tumors were found, but there were very few distant metastases. Nevertheless, vessels were often invaded by the neoplasms. Table 9.1 summarizes the results of these experiments.

9.4.3 *Discussion*

Lung cancers were induced in rats with exposures from 500 WLM to 14,000 WLM (Table 9.1a). These cancers appeared from the 12th to the 24th month after the beginning of exposure. The most efficient

TABLE 9.1a—*Lung tumors induced in rats by inhaled radon and radon daughters (CEN, Limoges, France)*

Exposure group	No. of rats	Median survival (days)	Animals with lung tumors			
			All lung tumors	Adenoma	Epidermoid carcinoma	Bronchiolo-alveolar adeno-carcinoma
Control	200	720	0			
21,000 WLM[a]	100	180	0			
14,000 WLM[b]	50	265	4	2	1	1
9600 WLM[c]	20	343	18	2	8	5 + 3[h]
7000 WLM[d]	50	485	15	6	4	5
4500 WLM[e]	40	659	11	2	4	5
3000 WLM[f]	40	714	5	1	2	2
500 WLM[g]	92	450	9	7	1	1

[a] Exposure regimen 4 hrs/day; 75 days. Radon decay product equilibrium 100%.
[b] Exposure regimen 4 hrs/day; 50 days. Radon decay product equilibrium 100%.
[c] Exposure regimen 5 hrs/day; 96 days. Radon decay product equilibrium 20%–30%.
[d] Exposure regimen 4 hrs/day; 25 days. Radon decay product equilibrium 100%.
[e] Exposure regimen 5 hrs/day; 60 days. Radon decay product equilibrium 20%–30%.
[f] Exposure regimen 5 hrs/day; 40 days. Radon decay product equilibrium 20%–30%.
[g] Exposure regimen 5 hrs/day; 115 days. Radon decay product equilibrium 1%.
[h] "Mixed" carcinoma.

cancer-producing exposure in these experiments ranged from 300 to 9000 WLM, delivered during 300 or 500 hours of exposure over a 3- or 4-month period. The optimum exposure rate was 200–750 WLM/week.

Above 14,000 WLM, benign tumors appeared; but since all animals died in the 12th month, cancers may not have had time enough to develop. The rats seldom died of infection or other cancerous complications, but rather of fibrosis of the interalveolar walls.

At lower exposure (500 WLM), the first cancers were found only in the 24th month. They appeared earlier as the exposure was increased. The earlier the first cancers appeared, the greater the number of cancers induced (over 50 percent of the lesions observed in some groups). Once a lesion was observed, its severity seemed to progress in relationship to the time after exposure. At lower doses, different lesions appeared distinctly in the following chronological order: metaplasia, adenomatous lesions, adenoma, and cancer. At higher exposures, this order was less evident, and all these types of lesions were found in the same lung. Mortality was concomitant with the appearance of tumors, at low exposures. Death occurred before the tumors appeared if the exposures were high, and the mortality rate increased as the periods of inhalation were lengthened. The authors thought that different equilibrium ratios of radon and radon daughters had no particular effect.

Stable cerium hydroxide was an efficient cofactor, and radon daughter induced cancers appeared two or three months earlier in those also exposed to the cerium hydroxide. Uranium ore dust *given on alternate days* did not influence the tumorigenesis.

TABLE 9.1b—*Dose effect relationship in male SPF rats with increasing radon daughter exposure (from Chameaud et al., 1981)*

Cumulative Exposure (WLM)	Number of Rats	Number of Lung Cancers	Lung Cancer Incidence per 10^6 per WLM
65	500	12	370
170	294	14	276
290	21	2	328
860	20	4	233
1470	20	5	170
3000	40	17	142
3800	20	7	92
3900	50	17	87
4500	40	29	161
6000	29	12	69
6000	25	11	73
8000	180	76	53

To determine whether radon daughters are effective in producing lung cancers at lower cumulative exposures than 500 WLM, another experiment was performed utilizing values to 65 WLM. These data are shown in Table 9.1b and indicate that lung cancer incidence does persist at the lower exposures.

An experiment using rats with tobacco smoke as a cofactor is reported (Chameaud *et al.*, 1978, 1981). At 12 months after the beginning of exposure, the time for the first appearance of cancers did not seem to be shortened, but data indicated that radiological doses necessary to produce pulmonary carcinoma may be lowered. In this study, exposure to cigarette smoke commenced after the radon daughter exposure. The total tumor yield was greater than in a group of rats exposed to radon and daughters alone. These data are shown in Table 9.1c.

9.5 Studies of the Biological Effects of Radon Daughters and Other Uranium Mine Air Contaminants at Battelle Pacific Northwest Laboratory

The past experiments at Battelle (Cross *et al.*, 1978; Stuart *et al.*, 1978) were designed to determine the roles of the many factors implicated in radiation carcinogenesis and other associated lung diseases in uranium miners. They involved life span exposures of beagle dogs, Syrian Golden hamsters and chronic exposure of rats to characterized aerosols of the potentially pathogenic air contaminants, singly and in combination.

9.5.1 *Studies Using the Syrian Golden Hamster*

The experimental design for the hamster studies is shown in Table 9.2. In a preliminary study, hamsters were exposed to levels of radon

TABLE 9.1c—*Effect of tobacco smoke on lung cancer induction in male SPF rat (from Chameaud et al., 1981)*

Cumulative Exposure (WLM)		Number of Rats	Number of Lung Cancers
4000	Radon daughters alone	50	17
4000	Radon daughters plus tobacco smoke	50	34
500	Radon daughters alone	28	2
500	Radon daughters plus tobacco smoke	30	8
100	Radon daughters alone	28	0
100	Radon daughters plus tobacco smoke	30	1
0	Tobacco smoke alone	45	0

TABLE 9.2—*Experimental design for hamster studies, Battelle Pacific Northwest Laboratory*

Group No.	Animals	Exposure Chamber Contents
1	102	Room air
2	102	Radon and radon daughters
3	102	Radon, radon daughters, and uranium ore dust
4	102	Uranium ore dust
5	102	Diesel engine exhaust
6	102	Diesel engine exhaust, radon, radon daughters and uranium ore dust

and radon daughters (with and without the presence of uranium ore dust) that produced a variety of pulmonary changes, including squamous metaplasia, with no life span shortening. For the studies described here, it was decided to raise the radon and radon daughter exposure levels in order to increase pulmonary pathology to a point where neoplasia might develop.

Methods

The six controlled-atmosphere test chambers used in this study were plexiglass spheres, 178 cm in diameter, with a volume of 3000 liters. Room air entered at the top of each sphere, was drawn through an opening at the bottom to a rotameter and flow control valve, and then exhausted through a filter. Radon was generated separately for each of the chambers of Groups 2, 3, and 6 by flowing compressed air through a prehumidifying buffer solution, a radium solution trap, the radium solution itself, a second trap, then a final membrane filter mounted at the chamber inlet. Radium concentrations were adjusted to maintain radon concentrations in the chambers between 0.21 and 0.25 microcuries/liter, at a flow rate through the chambers of 50 liters/minute. As shown in Table 9.3, this was sufficient to maintain the radon daughter concentration at a level of about 800 WL in the chambers for Groups 3 and 6. In the Group 2 chamber, which contained only room air aerosols and added radon, the unattached fraction of RaA was relatively high, causing a loss of radon daughters by diffusion to exposed surfaces. Radon levels were raised to concentrations between 0.26 and 0.31 nCi/L; radon daughter concentrations averaged somewhat less than 700 WL. Three chambers were equipped with dust feed aerosol generators, which added respirable carnotite ore dust to the inlet air stream.

Chambers of Groups 5 and 6 were equipped to supply diluted diesel engine exhaust to the incoming air. The source of this exhaust was a 3-cylinder, 43-bhp diesel engine, driving a 15-kilowatt generator connected to a series of resistance coils. Chamber air concentrations of carbon monoxide during each day's 6-hour exposure period were automatically controlled to 50 ± 3 ppm. The NO_2 content of the chambers ranged from 4 to 6 ppm, while SO_2 and aliphatic aldehyde levels remained below the 1 ppm detection level.

Sixteen months after the start of exposures, at least 60 hamsters from each of the six groups had died or had been sacrificed. At that time 54 replacement animals were added to each group. Nine of these replacement animals from each group were sacrificed after 1, 2, 4, 6, 8 and 11 months of exposure, respectively, for studies of pulmonary pathology without possible complication due to morbidity.

Results

Pulmonary neoplastic lesions were found in three hamsters of Group 2 and one hamster of Group 3. Table 9.4 shows the cumulative exposures for hamsters with lung tumors and for selected hamsters

TABLE 9.3—*Hamster exposure chamber parameters*[a]

Group & Chamber No.	RaA (nCi/L)	RaB (nCi/L)	RaC (nCi/L)	Working Level	Particulate Concentration (mg/m³)	Condensation Nuclei (10³/cm³)
1	—	—	—	—	—	3.9 ± 2.8
2	150 ± 60	70 ± 50	40 ± 30	690 ± 380	—	5.1 ± 4.0
3	150 ± 60	90 ± 40	50 ± 20	790 ± 330	22 ± 7	130 ± 50
4	—	—	—	—	19 ± 8	36 ± 24
5	—	—	—	—	7.3 ± 3.0	160 ± 60
6	160 ± 60	90 ± 40	50 ± 20	810 ± 310	23 ± 10	210 ± 90

[a] All values = mean ± S.D. of daily measurements.

TABLE 9.4—*Hamster exposure levels resulting in lung tumors or severe pulmonary epithelial metaplasia*[a]

Group	Exposure Rate (WL)	Cumulative Exposure (WLM)	Pulmonary Change
2	690 ± 380	8500 ± 4700	Squamous Carcinoma
2	690 ± 380	8500 ± 4700	Carcinoma *in situ*
2	690 ± 380	9100 ± 5000	Carcinoma *in situ*
2	690 ± 380	5100 ± 2800	Severe Squamous Metaplasia
2	690 ± 380	7100 ± 3900	Severe Squamous Metaplasia
3	790 ± 330	12,000 ± 5100	Squamous Carcinoma
3	790 ± 330	7200 ± 3000	Squamous Metaplasia
6	810 ± 310	9800 ± 3800	Severe Squamous Metaplasia

[a] All values = mean ± S.D.

with pulmonary lesions considered preneoplastic. All hamsters with lung tumors had cumulative exposures greater than 8000 WLM, while those with severe metaplastic lesions had received less exposure. The tumor in the lungs of one Group 3 hamster was very extensive, indicating that it had probably originated a considerable time before the death of the animal.

Within the first year of exposure, adenomatous proliferation (displacement of alveolar lining cells with ciliated cuboidal epithelium) of alveolar epithelium was significantly higher in all animals exposed to radon daughters (Groups 2, 3, and 6). In animals with longer exposure histories, those cells began to show vertical and horizontal crowding as they took on a columnar appearance. This was followed by squamous metaplasia of the cells, accompanied by a loss of cilia. In the lungs of many animals, all three stages of the progression could be seen in the same section.

The stage at which the lesions became frank carcinomas could not be ascertained, but a precancerous status is suspected for at least the later stages in this progression. Two of the squamous tumors found in the lungs of hamsters from Group 2 were diagnosed as carcinoma *in situ* according to cytological criteria, but no evidence of invasion of pulmonary vessels or of supportive structure could be found. An invasive squamous carcinoma was found in the lungs of one hamster of Group 2 and an epidermoid carcinoma was found in the lungs of one Group 3 hamster.

In Table 9.5, pulmonary lesions other than neoplasia, plus statistical analysis of this data, are listed by hamster exposure group. The animals that died in 1971 were exposed for 2–14 months, while those listed for 1972 had longer than 14 months exposure. Accumulation of particulate materials in the lungs was most pronounced throughout the experiment in those hamsters exposed to diesel engine exhaust.

The 54 replacement hamsters added to each group, which were periodically sacrificed in groups of nine for studies of pulmonary changes, allowed direct comparison of the pathogenesis of lesions in animals exposed to various agents for the same periods of time. The incidence of various pulmonary lesions is listed in Table 9.6 for each exposure group at the time of sacrifice. The most significant lesions discovered in Group 2 animals were the areas of alveolar epithelial bronchiolization in the lungs of two hamsters killed after 11 months of exposure. In both animals there was progression to squamous metaplasia. Squamous metaplasia of hyperplastic epithelium occurred in the lungs of 3 of 4 hamsters at 11 months in Group 3. The lungs of Group 4 animals were affected by lesions at much the same rate as that described for Group 3, particularly with regard to emphysema, septal cell hyperplasia, acute interstitial pneumonitis, and accumulation of pulmonary macrophages.

The most striking feature of the lungs of animals exposed to diesel engine exhaust (Group 5) was the heavy accumulation of soot over the longer exposure periods, both in alveolar macrophages and in alveolar air spaces; pulmonary emphysema was a significant finding. The lungs of Group 6 animals exhibited very similar lesions to those of Group 5, with slightly higher frequency of occurrence and slightly greater severity throughout the experiment; severe lesions of adenomatous proliferation of the alveolar epithelium were observed.

Discussion

Exposure to particulate matter, i.e., uranium ore dust and diesel exhaust soot, provoked inflammatory and proliferative responses in lungs consisting of macrophage accumulation, alveolar cell hyperplasia, and adenomatous alteration of alveolar epithelium. Additionally, exposure to radon + radon daughters was associated with increased occurrence of bronchiolar epithelial hyperplasia and with metaplastic changes of alveolar epithelium.

Squamous carcinoma developed in four hamsters, three of which received exposure to radon + radon daughters, and one to radon + radon daughters + uranium ore dust. Squamous carcinoma occurred only in association with squamous metaplasia of alveolar epithelium which, as stated above, occurred only in hamsters receiving exposure to radon + radon daughters alone or in combination with uranium ore dust. Several animals in Groups 2 and 3 developed squamous metaplasia of alveolar epithelium, but carcinoma was not diagnosed. Thus, it appeared that after exposure to radon + radon daughters, development of squamous metaplasia and development of carcinoma were related.

9.5.2 *Studies Using Beagle Dogs*

The experimental design for the dog studies is shown in Table 9.7. Beagle dogs were trained to accept daily smoking of 10 cigarettes (Groups 2 and 3) or to sham smoke unlighted cigarettes (Groups 1 and 4) for identical periods.

Methods

Two 10-dog exposure chambers were used in these studies. These chambers provided space for simultaneous head-only exposures of 20

TABLE 9.5—*Incidence of Respiratory Tract Lesions in Hamsters; Lifespan Studies*

Exposure Group:	1 (*Control*)	2 (*Radon daughters*)		3 (*Radon daughters + uranium ore*)		4 (*Uranium ore*)		5 (*Diesel exhaust*)		6 (*Radon daughters, uranium ore and diesel exhaust*)	
Year of Death:		1971	1972	1971	1972	1971	1972	1971	1972	1971	1972
No. of Animals:	82	51	45	55	46	50	49	55	40	57	49
A. Macrophage Accumulations											
1–2[a]	0	7	8	36	34	27	35	46	22	54	38
3–4[b]	0	1	12	2	8	1	6	3	17	3	8
More[c]	0	0	1	0	1	1	0	0	0	0	0
B. Emphysema											
1–2	5	16	10	17	19	9	18	17	5	11	12
3–4	2	2	8	2	15	0	9	3	7	2	13
More	0	0	0	0	0	0	0	0	1	0	0
C. Alveolar Septal Cell Hyperplasia											
1–2	1	9	10	9	26	11	21	23	12	20	26
3–4		0	12	1	6	0	4	2	5	1	8
More		0	1	0	1	0	0	0	0	0	0
D. Adenomatous Proliferation of Alveolar Epithelium											
1–2	2	8	11	12	20	2	11	6	11	13	23
3–4	0	3	13	3	15	2	6	0	3	0	11
More	0	0	1	0	0	0	0	0	0	0	0
E. Alveolar Epithelial Squamous Metaplasia	0	[2	7]	[1	12]	0	0	0	0	0	4
F. Bronchial or Bronchiolar Epithelial Hyperplasia											
1–2	1	19	14	22	30	11	18	11	15	26	25
3–4		0	15	2	3	1	9	3	6	0	14
More		0	1	0	0	0	0	0	0	0	0

G. Bronchiolar Epithelial Hyperplasia with Atypia	0	0	6	[2	6]	2	1	0	1	[1	10]
H. Bronchiolar Epithelial Hyperplasia with Squamous Metaplasia	0	[8	1]	6	0	1	0	0	0	[9	1]
I. Tracheal or Laryngeal Epithelial Hyperplasia	0	0	4	2	4	1	3	1	2	0	3
J. Bronchitis or Bronchiolitis	0	0	0	2	2	2	1	0	0	0	1
K. Laryngitis	0	0	5	[0	11]	1	5	1	4	1	1
L. Interstitial Pneumonitis											
1–2	12	5	3	10	19	8	21	2	12	6	14
3–4	9	3	8	1	4	0	4	3	6	1	10
More	0	0	0	0	0	0	0	0	0	0	0
M. Alveolar Septal Fibrosis (Pneumoconiosis)	0	0	0	0	0	[1	15]	0	1	[2	19]

[a] Very slight to slight.
[b] Moderate to marked.
[c] Severe.
Note: [ ] All severities of lesions combined are significantly different from controls (combined) at the 0.05 level of significance.

TABLE 9.6—*Incidence of non-neoplastic pulmonary lesions in hamsters added to the exposure groups and sacrificed periodically*

Exposure Group:	1 (Control)						2 (Radon daughters)						3 (Radon daughters + uranium ore)						4 (Uranium ore)						5 (Diesel exhaust)						6 (Radon daughters, uranium ore and diesel exhaust)					
No. of Months Exposure	1	2	4	6	8	11	1	2	4	6	8	11	1	2	4	6	8	11	1	2	4	6	8	11	1	2	4	6	8	11	1	2	4	6	8	11
No. of Animals																																				
Focal Accumulation of Macrophages	5	8	6	6	6	5	7	6	6	6	3	5	6	6	7	6	6	4	6	6	9	5	6	4	6	8	6	6	5	5	6	6	6	6	6	4
Emphysema	0	0	0	0	0	0	0	0	0	0	0	0	0	5	6	2	6	4	0	2	8	1	4	3	0	5	4	4	4	5	0	5	6	6	6	4
Interstitial Pneumonitis	0	0	0	1	1	0	0	0	0	0	0	0	0	1	3	3	2	2	0	1	7	0	2	0	0	2	3	2	4	5	0	2	4	4	6	4
Septal Cell Hyperplasia	1	0	0	0	1	2	0	0	1	1	1	2	0	2	1	1	3	3	0	0	9	0	3	3	0	0	4	0	1	3	0	2	4	2	5	4
Adenomatous Proliferation of Alveo-	1	0	0	0	1	0	0	0	0	0	0	0	0	1	4	0	0	1	1	1	5	0	0	0	0	4	1	2	2	2	2	3	5	4	3	2
lar Epithelium	0	1	0	2	2	0	2	3	0	0	0	2	0	2	1	1	3	4	1	3	5	1	1	3	1	5	4	4	3	3	3	3	3	4	5	3
With Squamous Metaplasia												2						3																		1
Bronchiolar Epithelial Hyperplasia	1	3	1	0	0	0	2	0	1	1	0	0	0	0	1	0	0	0	4	2	0	1	1	2	0	2	1	0	0	2	1	1	0	0	2	2

TABLE 9.7—*Experimental design for beagle dog studies*

Group No.	No. of Animals	Exposure Conditions
1	20	Radon, radon daughters, and uranium ore dust
2	20	Radon, radon daughters, uranium ore dust and cigarette smoking
3	20	Cigarette smoking
4	9	Room air (sham cigarette smoking)

dogs to radon daughters and carnotite ore dust. An aerosol diffusion system was used in each chamber in order to channel fresh aerosol past each dog's head. The uranium ore dust was added to the radon-laden air streams with dust feed mechanisms.

Each dog's general health was monitored daily by exposure technicians and animal care personnel; periodically each dog received a complete physical examination, including thoracic radiography. All animals were weighed biweekly; respiration rate, minute and tidal volumes were measured monthly; blood was sampled quarterly to permit hematology and clinical chemistry measurements, and occasionally for carboxyhemoglobin and plasma thiocyanate levels.

In the majority of cases, Groups 1 and 2 animals were sacrificed when death appeared imminent because of pulmonary insufficiency. Groups 3 and 4 animals were sacrificed at periods corresponding to high mortality peaks in Groups 1 and 2 so that tissues might be compared.

Results

During the 4½ years that Groups 1 and 2 dogs were exposed to radon daughters and uranium ore dust, radon daughter concentrations averaged 605 WL, with a standard deviation of 169 WL. During the same period of time, the average radon concentration was 105 ± 20 nCi/L; the uranium ore dust 12.9 ± 6.7 mg/m^3; and the condensation nuclei, 73,000 ± 20,000 per cm^3. Mass median aerodynamic diameters of the uranium ore dust ranged from 0.6 to 2.1 μm, with geometric standard deviations of 1.8 to 2.6. Activity median aerodynamic diameters, measured subsequent to exposure, averaged 0.60 μm, with a geometric standard deviation of 1.7. Unattached RaA averaged 3 percent. Unattached percentages of RaB and RaC were only small fractions of RaA levels.

Tables 9.8 and 9.9 show the exposure history of Groups 1 and 2 dogs up to the time of their deaths. Cumulative exposures (WLM) were

TABLE 9.8—*Cumulative exposure of group 1 dogs; radon daughters with uranium ore dust*

Dog No.	Months of Exposure	Cumulative[a] Exposure (WLM)	Neoplasms
610	34	9,410	
505	37	10,400	
509	39	11,400	
520	42	10,800	
577	43	11,000	
642	44	12,800	
615	46	13,300	Nasal carcinoma
539	48	11,900	
608	48	14,000	Epidermoid carcinoma and nasal carcinoma
524	51	13,100	
512	51	14,200	Bronchioloalveolar carcinoma
567	52	14,900	
514	54	15,700	
522	54	15,600	
523	54	15,300	Epidermoid carcinoma
531	54	15,700	Bronchioloalveolar carcinoma
541	54	15,700	Epidermoid carcinoma
540	54	15,700	Fibrosarcoma
525	54	15,700	Bronchioloalveolar carcinoma

[a] At time of death (one animal sacrificed for study of early pathological change).

calculated by multiplying the actual number of hours each dog was exposed by the average radon daughter concentrations (605 WL) and dividing by 170. Some animals had equivalent or higher numbers of exposure months but lower cumulative radon daughter exposure than others because they were removed from the experiment for a period of time due to illness. No significant (at the 0.05 level of significance) differences between survival curves were noted for these animals. There were no significant differences between experimental and control mean values of body weights and clinical chemistry parameters. However, Groups 1 and 2 dogs showed significant increases in neutrophil levels over controls. Respiratory rates increased and minute volumes decreased in Groups 1 and 2 animals compared to controls.

Minimal pathologic changes occurred in the respiratory tracts of the sham-exposed control dogs. These included very slight basal cell hyperplasia in both the tracheal and laryngeal mucosa, as well as slight glandular hyperplasia in these organs in dogs that received up to 65 months of sham exposures.

TABLE 9.9—*Cumulative exposure of group 2 dogs; radon daughters with uranium ore dust plus cigarette smoking*

Dog. No.	Months of Exposure	Cumulative[a] Exposure (WLM)	Neoplasms
855	33	9,780	
591	37	10,200	
629	38	9,240	
585	42	11,200	
595	42	11,500	
552	44	12,700	
587	46	11,800	
627	46	13,000	
516	46	12,200	
562	46	12,900	
637	46	13,200	
504	47	13,200	
593	50	14,300	
544	51	13,800	Nasal carcinoma
573	52	14,900	
545	52	15,100	
551	52	15,000	
530	52	15,100	
518	52	12,000	Bronchioloalveolar carcinoma

[a] At time of death (one animal sacrificed for study of early pathological change).

Group 1 (Radon Daughters and Uranium Ore Dust) and Group 2 (Radon Daughters, Uranium Ore Dust, and Cigarette Smoke).

The earliest histological manifestation of exposure to radon daughters and uranium ore dust was a brown pigment, considered to be ore dust, in pulmonary macrophages. Tracheobronchial lymph nodes contained large amounts of uranium ore dust. Pulmonary hyalinosis was a common microscopic change in Group 1 and Group 2 dogs that died or were sacrificed after exposure for 27 months or longer. Vesicular emphysema was present in the lungs of all dogs from Groups 1 and 2. The lesions were severe in dogs that had been exposed to radon daughters and uranium ore dust for 30 to 45 months; and, in dogs with exposure histories longer than 45 months, bullous emphysema was common.

Pulmonary fibrosis was prevalent in all dogs from Groups 1 and 2. Alveolar septal fibrosis was apparent to a slight degree in two dogs sacrificed after only six months exposure, and was progressively worse after longer exposure. Pleural thickening due to fibrosis was consistently severe in dogs exposed to radon daughters and uranium ore dust.

Alveolar epithelial changes were prominent in the lungs of dogs from Groups 1 and 2 that were exposed for longer than 30 months. In dogs with exposure histories of 40 months or more, large areas of adenomatosis were present. Frequently, the lesion had progressed to squamous metaplasia of the alveolar epithelium, in which atypical cells were present. After approximately 50 months of exposure to radon daughters and uranium ore dust, lungs from 11 of 21 Groups 1 and 2 dogs that had died or were killed contained large cavities within the parenchyma. Each cavity was surrounded by hyperplastic adenomatous epithelial cells.

The development of neoplasms is shown in Tables 9.8 and 9.9. In two of the Group 1 dogs, #523 and #608, epidermoid carcinomas were associated with the cavities described above. The tumors were solitary masses in peripheral areas of single lobes and consisted of irregularly shaped lobules of non-keratinizing stratified squamous epithelium with a scant amount of connective tissue stroma. They had locally invaded adjacent alveoli. Another epidermoid carcinoma was a well-circumscribed mass composed of anaplastic cells and numerous mitotic figures. Three other primary lung tumors, bronchioloalveolar carcinomas were present in dogs #512, #525, and #531. The tumors were solitary masses associated with distal bronchioles in single lobes; had locally invaded adjacent alveoli; and had not metastasized. A bronchioloalveolar adenoma, composed of a papillary proliferation of epithelium, as well as an epidermoid carcinoma was present in Group 1 dog #541. A fibrosarcoma was present in the right apical lobe of the lung from Group 1 dog #540.

Squamous cell carcinomas arising from the epithelium of the nasal cavity occurred in three dogs. Histologically, the nasal carcinomas were composed of non-keratinizing stratified squamous epithelium forming irregular cords infiltrating the submucosa and extending between spicules of bone. A large metastasis was present in the mandibular lymph node of dog #608. Dog #608 also had an epidermoid carcinoma in the lungs, not necessarily metastasized from the nasal carcinoma.

Group 3 (Cigarette Smoke Only).

Three dogs from Group 3 were sacrificed after 65 months of exposure to cigarette smoke. Mild chronic inflammatory changes in the lungs were observed. These dogs had increased amounts of phagocytized exogenous yellow pigment (considered to be associated with cigarette smoking), found primarily in peribronchiolar and perivascular areas. There was more subpleural vesicular emphysema in the smoke-exposed dogs than in the sham-exposed control dogs. The tracheas of two smoke-exposed dogs and the larynx of one smoke-exposed dog had thickened basement membranes.

The tracheobronchial lymph nodes of all smoke-exposed dogs contained large amounts of the same yellow pigment within the histiocytes, and had lesions clearly related to exposure. Moderate reactive lymphoid hyperplasia in the tracheobronchial lymph nodes and the moderate histiocytosis in the mediastinal lymph nodes were probably related to the observed chronic inflammatory changes in the lung of one smoke-exposed dog.

Discussion

From examination of the mortality data in Tables 9.8 and 9.9, it is evident that all animals with respiratory tract tumors (except dog #518) had cumulative exposures greater than 13,000 WLM. (That dog was also unique in that it survived for 28 months after completion of exposures to radon daughters and uranium ore dust.)

Slight differences occurred between Groups 1 and 2 dogs that could be attributed to smoke exposure. Generally, the incidence and degree of severity of bullous emphysema were greater in the smoke-exposed dogs (Group 2). The onset of this condition was also noted earlier in the dogs of Group 2 than in those of Group 1.

9.5.3 *Studies Using SPF Rats and Hamsters*

Several studies are underway to determine the responsible agents (and their interactions) that may cause bronchogenic lung cancer in uranium miners. These include studies that examine the influence of altered radiological dose rate and the physiologic or pathogenic role of uranium ore dust in neoplastic response.

Investigations of altered pathogenic response caused by inhaled radon daughters as a function of dose rate and total dose involve daily inhalation exposure of groups of male SPF Wistar rats to several concentrations. In one experiment, rats and hamsters were exposed simultaneously in groups of 32 and 34, respectively, for 84 hours per week to approximately 900 and 400 Working Levels of radon daughters, with and without 15 mg/m^3 of carnotite uranium ore dust. Controls were housed in identical cages and chambers, but exposed to room air only. Exposures were stopped after 5 months, at which time the animals had received cumulative exposures to radon daughters of approximately 9000 WLM with ore dust and 4000 WLM without ore dust.

The animals were then held for the duration of their life spans and results were compared with those from previous experiments in which hamsters were exposed for their full lifetimes to similar atmospheres and cumulative exposures, but at approximately one-third the exposure rate. Exposure to radon daughters with or without added ore dust had no influence on the survival rate or life span of the hamsters; however, exposures of rats to radon daughters, with or without uranium ore dust, decreased their survival time.

Although hamsters exposed to radon daughters with dust in the previous experiments had a 1.0 percent incidence of pulmonary squamous carcinoma (see Section 9.5.1), recent findings (Table 9.10) with comparable exposures delivered to rats at three times the rate, showed 60 percent incidence of squamous carcinoma and adenocarcinoma. While there may exist a significant difference between the two rodent species in susceptibility to induction of pulmonary carcinoma (60 percent for rats vs. 1 percent for hamsters), these findings suggest the possibility of a dose-rate effect in radon daughter-induced pulmonary carcinogenesis. Further studies are in progress to test this hypothesis.

9.6 Discussion of Studies of Inhaled Radon Daughters in Experimental Animals

The studies described above concerning the biological effects of inhaled radon and radon daughters in experimental animals were undertaken to determine dose-response relationships under controlled exposure conditions and to define the correlation between exposure and absorbed dose in respiratory tissue.

In the University of Rochester studies, mice received cumulative inhalation exposures of 14,000–72,000 WLM of radon daughters without added dust over 10–35 weeks, dogs received 200–10,000 WLM over 1–50 days, and rats received 25,800 WLM over 24 weeks. Destructive, hyperplastic, and metaplastic lesions appeared in the lungs of the three species. Adenomas and "tumorlets" appeared in mouse lungs and chronic inflammation appeared in dog lungs, but no frank carcinoma. Tumorlets, two adenomas, and two carcinomas were found late post-exposure in rat lungs.

In the CEN studies, rats developed lung cancer between 12 and 24 months after having received 3 to 4 months of daily 4- to 5-hour exposures to radon daughters, with and without uranium ore dust, totaling 500 to 14,000 WLM. Observed pulmonary cancers included epidermoid (squamous) carcinoma, bronchiolar-adenocarcinoma, bronchioalveolar adenocarcinoma, and "mixed" carcinoma. At 9600 WLM, cancer incidence approached 90 percent.

In the Battelle Pacific Northwest Laboratory studies, daily life span inhalation exposures of rodents and dogs to 600–800 WL of radon daughters with uranium ore dust at about 15 mg/m^3 produced pulmonary fibrosis, emphysema and neoplastic lesions of the respiratory tract. Findings included squamous carcinoma and bronchioalveolar carcinoma in 25 percent of the dogs receiving 4 to 4½ years of daily 4-hour exposures to radon daughters with uranium ore dust at cumulative radon daughter exposures of 12,000 to 16,000 WLM. In addition, SPF rats exposed to 9,200 WLM of radon daughters with uranium ore dust have shown squamous cell carcinoma, adenocarcinoma, and mixed carcinoma in 60 percent of those receiving about 5½ months of daily exposures.

Four variables emerge as affecting the outcome of tumor location or response in the studies described above using experimental animals, whose goal is extrapolation of radon/radon daughter carcinogenesis relationships to man. These are: (1) *absorbed dose* (rad per WLM) to critical respiratory tract tissue; (2) *role of concomitant dust* and resulting radon daughter attachment fractions on the site of deposition and distribution of alpha irradiation; (3) *total exposure* as Working Level Months; and (4) *exposure rate* as WLM per week. Absorbed doses to all basal cells of the hamster airway epithelium were calculated by Desrosiers *et al.* (1978) as 0.1–0.3 rad/WLM, with 0.1 rad/WLM to "4th airway group basal cells" and 0.4 to 1.2 rad/WLM for the finest bronchioles. However, these calculations did not account for nasal deposition. Morken measured levels of 0.17 rad/WLM to the lungs from mice and dogs by dissecting and counting whole lungs; doses to the bronchial tissue of the mouse may have been 5–10 times the mean lung dose and measured doses to the dog trachea were 30 times this level. No absorbed doses were measured in the studies in France. At Battelle Pacific Northwest Laboratory, preliminary measurements of absorbed dose indicated 0.2–0.3 rad/WLM in hamster lungs, with 2–3 times higher levels in the trachea. These figures are, at first glance, fairly consistent, and the values for rodents are considerably lower than those calculated for the subsegmental bronchial epithelium in man for the same reference atmosphere. However, this does not necessarily suggest that correspondingly higher exposures are necessary in animals to study radon daughter carcinogenesis, but rather reflect the different sites affected in animal lung versus human lung.

The role of concomitant exposure to uranium ore dust is currently under study at the Battelle Pacific Northwest Laboratory. Its absence causes greater unattached radon daughter fractions (about 20 percent

TABLE 9.10—*Incidence of lesions in hamsters and rats following exposure to radon daughters with and without uranium ore dust*

	Exposure Conditions	Results Following Exposures[a]		
		Nasopharynx	Trachea	Lung
Group 1	Laboratory Air Controls			
34 Hamsters		N	N	N
32 Rats		N	N	N
Group 2	375 WL Radon Daughters Alone, 3800 WLM (unattached fraction 16–20%)			
		34 Squamous Metaplasia	N	31 Slight Bronchiolization
34 Hamsters		1 Squamous Carcinoma	5 Slight Hyperplasia	17 Slight Radiation Pneumonitis
32 Rats		32 Squamous Metaplasia	7 Squamous Metaplasia	32 Moderate Bronchiolization
		2 Squamous Carcinoma		16 Moderate Radiation Pneumonitis
				10 Adenomatosis
				2 Squamous carcinoma
				1 Squamous Carcinoma & Bronchioloalveolar carcinoma

Group 3	900 WL Radon Daughters with Uranium Ore Dust (Carnotite, 15 mg/m^3) 9200 WLM (unattached fraction 1–2%)			
34 Hamsters		14 Very Slight Squamous Metaplasia	N	32 Slight-Moderate Bronchiolization 2 Extensive Fibrosis Adenomatosis Emphysema
32 Rats		26 Squamous Metaplasia	N	17 Squamous Carcinoma 1 Adenocarcinoma 1 Squamous Carcinoma & Adenocarcinoma

[a] N = normal.

vs. 1–3 percent) in these experiments and this may be a factor in the higher proportion of nasal carcinoma seen at that laboratory in hamsters and rats, as shown in Table 9.10. The presence of fibrogenic ore dust may impede clearance of deposited radon daughters or may set up regions of damage in the lungs that could promote carcinogenesis. However, the studies at the CEN of pulmonary carcinogenesis in rats show a somewhat higher total incidence of lung cancer after 9600 WLM of radon daughter exposure without ore dust than the data of Table 9.10. Thus, the role of ore dust is still unresolved.

The data from all experiments suggest that exposure rate influences tumor production. Tumor probability increases with cumulative radon daughter exposure similar to the findings at the Battelle Pacific Northwest Laboratory.

Studies with mice and rats at the University of Rochester involved exposure rates of 1650 WLM/week and 1100 WLM/week, respectively; few carcinomas were produced and no exposure-effect relationships were developed. Conversely, experiments with rats at Battelle Pacific Northwest Laboratories showed 60 percent pulmonary carcinoma incidence at 9200 WLM delivered at a rate of 450 WLM/week. This exposure rate was similar to the most efficient carcinogenesis rate of 200–750 WLM/week described in the studies at the CEN.

These findings suggest that two possibilities remain which may influence the extrapolation of animal data to man.

a. The presence of the ore dust may alter pulmonary pathogenesis.
b. The rate of radon daughter exposure may influence tumor probability.

Both possibilities could cast doubt upon the adequacy of the cumulative working level month concept as a sufficient index of carcinogenic or other lethal pulmonary disease risk, if exposure conditions differ radically.

10. Calculated Lung Cancer Risk to Individuals from Radon Daughter Exposure

10.1 Introduction

It is important to use the knowledge gained through the uranium mining experience to predict lung cancer risk which may arise either normally, through a particular environmental practice or occupationally. Radon daughters are ubiquitous and elevated exposures of individuals or relatively large groups are possible. It is desirable to quantitate their risk for comparison with that expected in normal or average environments. Elevated radon daughter exposures are now reported in association with ordinary living conditions as well as unusual geological settings or occupational settings. Some of these are: homes that are poorly ventilated (especially single family dwellings where living space is close to the soil), basements, crawl spaces with exposed soil, areas adjacent to or homes built upon uranium mill tailings, homes that are supplied with radon-rich water, and homes near phosphate-rich areas or near phosphate tailings piles, to name but a few. As more measurements of natural radioactivity are performed, it will be necessary to establish whether specific situations are tolerable with regard to lung cancer risk. The following model has been developed so that individual exposures may be assessed.

10.2 Predictive Model

Bronchogenic lung cancer induction by long exposure to elevated levels (a few hundred working level months or more) of radon daughters in underground mines is well established (Lunden *et al.*, 1971; Snihs, 1973; Sevc *et al.*, 1976; Kunz and Sevc, 1978; Axelson and Sundell, 1978; Archer *et al.*, 1979; Kunz *et al.*, 1979; NAS, 1980;

Waxweiler, 1981). Based upon the exposures described in Section 5, normal average environmental exposures could be near 0.2 WLM per year, resulting in a lifetime average exposure of $85 \times 0.2 = 17$ WLM. Although Snihs (1973) considers that the lowest underground exposure which resulted in an apparent increase in lung cancer deaths is about 15 WLM, epidemiological studies in other underground mining groups do not indicate a statistically significant excess of lung cancer below cumulative exposures of about 100 WLM. Some argue that since lung cancer mortality in miners at these low levels of exposure is not significantly different from expected values (Myers and Stewart, 1979), a threshold for radon daughter-induced lung cancer could exist. Hewett (1979) concludes from the analysis of Canadian uranium miners that if a threshold exists, it is below 60 WLM. Thus, although the possibility exists that environmental or slightly elevated radon daughter levels do not induce lung cancer, conservative radiation protection practice takes the view that radiation-induced lung cancer is a stochastic process with no threshold and this view is adopted in this report. The data from the higher mine exposures must be used to estimate possible lung cancer rates at low radon daughter levels, but the temporal conditions for mining versus environmental exposure (duration and age at first exposure) make it difficult to relate the two directly.

In spite of the difficulties, extrapolation from the higher exposures is the only method available and this approach has been taken in this report and by others (Stranden, 1980; Cohen and Cohen, 1980; Evans *et al.*, 1981; Harley and Pasternack, 1981). Spontaneous lung cancer mortality (nonsmoking related) offers some guidance since any risk model should not predict a lung cancer incidence that is greater than observed for ordinary background levels of radon daughters. The calculation adopted here is relatively simple and yields results that are consistent with both the underground mining experience and expected background incidence. The model is based upon the information about lung cancer enumerated below which appears reasonably certain. The confounding effect of smoking is considered later.

1. The highest reported rate of appearance of lung cancer attributable to radon daughters is 45×10^{-6} per year per WLM for older miners (see Section 8). The average value of this risk coefficient (10×10^{-6} per WLM) is adopted in this report and was obtained by using all groups where sufficient data were available (see Section 8).
2. The rate of appearance of lung cancer (following a latency interval) after a single external radiation exposure seems reasonably uniform with time, that is, there is no wave of lung cancer.

Support for this comes from the Japanese A-bomb data (Beebe *et al.*, 1978; Kato and Schull, 1982).

3. The appearance rate for a single exposure is highest when age at exposure is highest (Beebe *et al.*, 1978; Kato and Schull, 1982). This is also seen in the Czechoslovakian mining data following exposure to radon daughters over an extended period (Sevc *et al.*, 1976) (see Table 8.1).
4. The incidence of lung cancer before the age of 40 is rare (Saccomanno *et al.*, 1974; Israel and Chachinian, 1976).
5. The median age associated with lung cancer appearance in miners is about 60 in nonsmokers and 50+ in smokers, regardless of the age at first start of mining (Archer *et al.*, 1979, 1981; Saccomanno, 1981).
6. Radon daughter-induced lung cancer rarely, if ever, appears at less than five to seven years after exposure (Archer *et al.*, 1979, 1981).
7. The time for tumor growth from bizarre cells to frank appearance is about five years (Saccomanno *et al.*, 1974).

Reissland *et al.* (1976) originally proposed a model specifically for the appearance of leukemia in a population occupationally exposed to chronic external radiation. The annual appearance rate of tumors attributable to a single exposure was assumed constant and commenced after a constant latent interval. The modified absolute risk model for lung cancer adopted here is based on this technique but differs from it in two ways: (1), the incidence does not manifest itself until age 40 regardless of the age at exposure, and, after age 40, a minimum single value for the latent interval of 5 years applies; (2), the tumor rate is not uniform with time but is decreased from the time of exposure by an exponential factor with an effective half-life of 20 years. This exponential factor accommodates the age dependent incidence rates.

The annual appearance rates following a single exposure, at age 20 or at age 45, are shown schematically in Fig. 10.1. The uncorrected rates are shown in Curves (a) and (b), and the rates corrected by the exponential factor by Curves (c) and (d).

The exponential factor is justified by assuming that a loss half-time exists for stem cells that are transformed by alpha radiation (Harley and Pasternack, 1981). Using this approach, and correcting the attributable risk for each year subsequent to exposure with a 20-year half-life, the calculated lifetime risks agree well with those observed in the uranium mining studies, including the fact that miners having short-term exposures and first exposed at age 40 have a higher lifetime lung

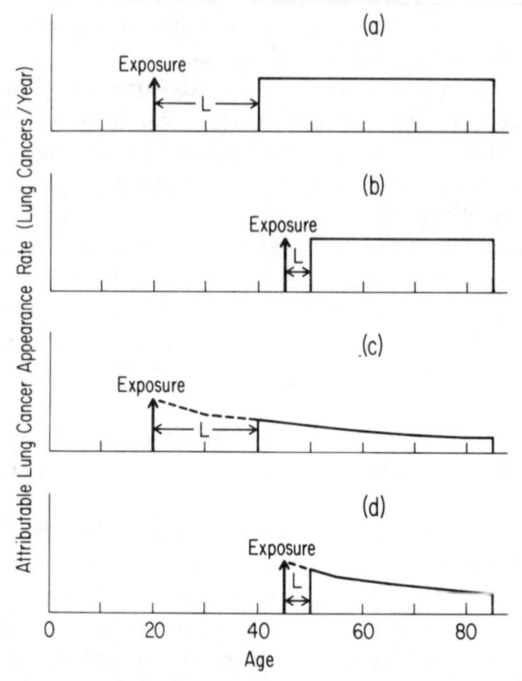

Figure 10.1. Models for annual appearance of lung cancer attributable to a single exposure of radon daughters at age 20 (a and c) or age 45 (b and d). Figures (a) and (b) not corrected for cell loss; in (c) and (d) a 20-year half-life is introduced to accommodate cell loss. L is the latent interval.

cancer risk than those first exposed at age 20. The calculated lifetime risk of lung cancer is also corrected by an appropriate life table value to account for the slight reduction in lung cancers due to death from other causes.

The numerical calculations to determine lifetime risk are carried out in the following manner. First the annual attributable risk is determined, subsequent to an annual exposure of 1 WLM at t_0:

$$A(t \mid t_0) = RC \left(\frac{\mathrm{P_t}}{\mathrm{P_{t_0}}}\right) \mathrm{e}^{-\lambda(t-t_0)} \tag{10-1}$$

$A(t \mid t_0)$ = attributable annual tumor rate at age t ($t \geqslant 40$) due to a single annual exposure at t_0. If exposure occurs after age 40, risk commences at t_0 +5 years, if exposure occurs before age 35, risk commences at age 40,

RC = risk coefficient 10×10^{-6} per year per WLM is adopted in this report,

$\mathrm{P_t}/\mathrm{P_{t_0}}$ = life-table correction to account for death from other causes

P_{t_0} is the probability that an individual will be alive at age t_0 and P_t is the probability that an individual will be alive at age t, and

λ = decrease in rate of risk expression due to repair, cell death or unspecified mechanisms ($\lambda = \frac{\ln 2}{20}$ yr^{-1}).

The lifetime risk, $LR(t_0)$, from this single annual exposure at t_0 is calculated by summing the annual attributable rate over the ages of tumor appearance

$$LR(t_0) = \sum_{t}^{85} A(t \mid t_0) \qquad \begin{array}{l} t = 40 \text{ to } 85 \text{ for } t_0 \leqslant 35 \\ t = (t_0 + 5) \text{ to } 85 \text{ for } t_0 > 35 \end{array} \qquad (10\text{-}2)$$

The lifetime risk for multiple (annual) exposures. LR, is obtained by summing lifetime risk from each single exposure, $LR\ (t)$, at t_0, t_1, . . . t_n:

$$LR = \sum_{t_0}^{t_m} LR(t) \qquad (10\text{-}3)$$

10.3 Lifetime Lung Cancer Risks from Model Predictions

The basic incidence data from the underground mining epidemiology studies cannot be applied directly to environmental situations since patterns of exposure differ. The common factor, however, should exist in the risk per rad for bronchial dose. The lifetime lung cancer risk attributable to an absorbed dose of 1 rad per year using the conversion factor of 0.5 rad/WLM estimated for miners, and the model described in Section 10.2, have been calculated. Table 10.1 shows the results of these calculations for exposures beginning at age 1, 10, 20, 30, 40, 50, 60 and 70 years and for exposure durations of 1, 5, 10, 30 years as well as lifetime exposure (to age 85 years). Since exposure of a population would involve persons of various ages, it is also of interest to know the lifetime risk of radon daughter induced lung cancer for a population with age characteristics typical of the United States. This is shown in the last column of Table 10.1, using the 1975 age distribution for the U.S. (WHO, 1978).

In Table 10.1, the risk as a function of basal cell dose (rad/year) from radon daughters is given. This is suitable for calculating lung cancer risk from either environmental or occupational exposures. Exposures acquired in underground mines are typically given in WLM. If exposure duration in years is known, an average value of WLM per year may be obtained and converted to absorbed dose rate in rads/year by multiplying by the factor 0.5 rad per WLM.

TABLE 10.1—*Lifetime lung cancer risk per rad per year from radon daughter exposure in mines. Lifetime risk as a function of age and duration of exposure*

Exposure Duration	Lifetime Lung Cancer Risk								Lung Cancers in a Population of 10^5 Persons[a]
	Age at First Exposure								
	1	10	20	30	40	50	60	70	
1 Year	9.2×10^{-5}	1.3×10^{-4}	1.8×10^{-4}	2.6×10^{-4}	3.0×10^{-4}	2.4×10^{-4}	1.8×10^{-4}	1.0×10^{-4}	19
5 Years	4.8×10^{-4}	7.2×10^{-4}	9.8×10^{-4}	1.4×10^{-3}	1.5×10^{-3}	1.2×10^{-3}	7.8×10^{-4}	4.0×10^{-4}	94
10 Years	1.1×10^{-3}	1.6×10^{-3}	2.2×10^{-3}	3.0×10^{-3}	2.8×10^{-3}	2.0×10^{-3}	1.3×10^{-3}	5.4×10^{-4}	190
30 Years	4.8×10^{-3}	6.8×10^{-3}	7.8×10^{-3}	7.8×10^{-3}	6.0×10^{-3}	3.6×10^{-3}	1.8×10^{-3}	5.4×10^{-4}	540
Life	1.3×10^{-2}	1.3×10^{-2}	1.1×10^{-2}	1.1×10^{-2}	6.4×10^{-3}	3.8×10^{-3}	1.8×10^{-3}	5.4×10^{-4}	800

[a] For a population with age characteristics equal to that in the whole United States in 1975.

Risk from a bronchial dose in rad per year to basal cells as shown in Table 10.1 is one way to evaluate enviromental exposures. Two other lifetime risk tables are derived from Table 10.1 that relate risk to environmental exposure in WLM per year, and an annual exposure to a radon concentration of 1 pCi $^{222}Rn/m^3$.

It was shown in Section 7 that the average environmental dose conversion factors for the adult male, female, ten-year-old child, and infant are 0.7, 0.6, 1.2, and 0.6 rad/WLM, respectively. The differences reflect reduced breathing rates under normal environmental conditions, different lung morphometry, and the increased percentage of unattached RaA in ordinary atmospheres. Thus, an environmental exposure is expected to be somewhat more productive of tumors than an equivalent occupational exposure in WLM. The system can be simplified considerably if we accept the environmental dose conversion factor of 0.7 rad/WLM which applies to adult males, for all people. The lifetime risk estimate which includes the effect of the higher dose conversion factor during the years of childhood is within 10 percent of this value. The lifetime risk per environmental WLM per year are shown in Table 10.2 for the conditions given with Table 10.1.

For the case of exposure measured as radon concentration and time, the average annual bronchial dose to adult males from the daughters associated with 1 pCi $^{222}Rn/m^3$ is obtained by combining Equations (5-8) and (5-9),

$$\begin{aligned} &\text{Dose (rad/year per pCi } ^{222}\text{Rn/m}^3) = \\ &0.0003 \text{ (hours per day active/24)} + \\ &0.0002 \text{ (hours per day resting/24)} \end{aligned} \qquad (10\text{-}4)$$

Assuming 16 hours per day are active and 8 hours per day are spent resting,

$$\text{Dose} = 0.00027 \text{ rad/year per pCi } ^{222}\text{Rn/m}^3 \qquad (10\text{-}5)$$

As with the WLM calculations, the value for adult males will be within 10 percent of that for all people. Table 10.3 shows the lifetime risks for annual exposures to 1 pCi $^{222}Rn/m^3$.

The lifetime risk of lung cancer estimated in this report ($1 - 2 \times 10^{-4}$ per WLM, dependent upon age and duration of exposure) may be compared with the risk as estimated by others. ICRP (ICRP, 1981) has adopted $1.5 - 4.5 \times 10^{-4}$ as the risk of lung cancer per WLM, based primarily upon the Czechoslovakian underground mining data. Evans *et al.* (1981) estimated an upper estimate for lifetime risk to be 1.0×10^{-4} per WLM for members of the general population from the U.S. and Czechoslovakian miner epidemiology and a consideration of

TABLE 10.2—*Lifetime lung cancer risk under environmental conditions per WLM per year.*[a] *Lifetime risk as a function of age and duration of exposure*

Exposure Duration	Lifetime Lung Cancer Risk								Lung Cancers in a Population of 10^5 Persons[b]
	Age at First Exposure								
	1	10	20	30	40	50	60	70	
1 Year	6.4×10^{-5}	9.1×10^{-5}	1.3×10^{-4}	1.8×10^{-4}	2.1×10^{-4}	1.7×10^{-4}	1.3×10^{-4}	7.0×10^{-5}	13
5 Years	3.4×10^{-4}	5.0×10^{-4}	6.9×10^{-4}	9.8×10^{-4}	1.0×10^{-3}	8.4×10^{-4}	5.5×10^{-4}	2.8×10^{-4}	66
10 Years	7.7×10^{-4}	1.1×10^{-3}	1.5×10^{-3}	2.1×10^{-3}	2.0×10^{-3}	1.4×10^{-3}	9.1×10^{-4}	3.8×10^{-4}	130
30 Years	3.4×10^{-3}	4.8×10^{-3}	5.5×10^{-3}	5.5×10^{-3}	4.2×10^{-3}	2.5×10^{-3}	1.3×10^{-3}	3.8×10^{-4}	380
Life	9.1×10^{-3}	9.1×10^{-3}	7.7×10^{-3}	7.7×10^{-3}	4.5×10^{-3}	2.7×10^{-3}	1.3×10^{-3}	3.8×10^{-4}	560

[a] For radon daughters measured under environmental rather than underground mining conditions.

[b] For a population with age characteristics equal to that in the whole United States in 1975.

the spontaneous lung cancer incidence. UNSCEAR (UNSCEAR, 1977) has reviewed the epidemiology in the uranium miners in Canada, the U.S. and Czechoslovakia, Swedish non-uranium miners and iron miners in the United Kingdom. UNSCEAR indicates that a lifetime lung cancer risk of $2 - 4.5 \times 10^{-4}$ per WLM can be regarded as probable. The Committee on the Biological Effects of Ionizing Radiation (NAS, 1980) reviewed lung cancer in U.S., Canadian and Czechoslovakian uranium miners, Newfoundland fluorspar miners and Swedish metal miners. The range of risk for all groups was expressed as a rate ($6 - 47 \times 10^{-6}$ per person per year per WLM) rather than lifetime risk. If we assume that attributable lung cancer expression takes place in the groups selected over a 30-year interval, then $6 - 47 \times 10^{-6}$ per person year per WLM is equivalent to a range of lifetime risk of about $2 - 14 \times 10^{-4}$ per WLM.

The lifetime risk estimates for lung cancer attributable to radon daughter exposure per WLM are, therefore, reasonably consistent considering the difficulty in estimating this quantity without complete follow up and the methodological problems indicated in Section 8. Tables 10.1 to 10.3 yield lung cancer risk estimates comparable to other values, but allow calculation of risk for various ages and exposure durations.

As an example, using Table 10.2, the lifetime risk of lung cancer for an average environmental exposure of 0.2 WLM/year would be (0.2) (0.0091) or 0.18%. This is one-fifth of the lifetime risk of lung cancer for nonsmokers of about 1 percent. Assuming that these lung cancers are expressed uniformly over a 45 year interval, this would amount to about 9000 lung cancer deaths per year in the U.S. population.

10.4 Relative and Modified Absolute Risk Projection Models

A relative risk model may also be used in estimating lifetime risk of lung cancer attributable to radon daughter exposure (constant increase per WLM over the normal lung cancer appearance rate). Whittemore and McMillan (1983) have reanalyzed the lung cancer mortality among the U.S. cohort of white underground uranium miners based on follow-up through December 31, 1977. Relative risk functions were examined using the proportional hazards model. They found that age specific lung cancer mortality among miners could be expressed as that for nonsmoking men born at the same time multiplied by a risk factor, R. The relative risk function giving the highest likelihood was,

TABLE 10.3—*Lifetime lung cancer risk under environmental conditions*[a] *per pCi* $^{222}Rn/m^3$. *Lifetime risk as a function of age and duration of exposure*

Exposure Duration	Lifetime Lung Cancer Risk								Lung Cancers in a Population of 10^5 Persons[b]
	Age at First Exposure								
	1	10	20	30	40	50	60	70	
1 Year	2.5×10^{-8}	3.6×10^{-8}	5.0×10^{-8}	7.1×10^{-8}	8.3×10^{-8}	6.7×10^{-8}	4.8×10^{-8}	2.7×10^{-8}	0.0051
5 Years	1.3×10^{-7}	1.9×10^{-7}	2.7×10^{-7}	3.8×10^{-7}	4.0×10^{-7}	3.1×10^{-7}	2.1×10^{-7}	1.1×10^{-7}	0.026
10 Years	2.9×10^{-7}	4.2×10^{-7}	5.8×10^{-7}	8.1×10^{-7}	7.5×10^{-7}	5.6×10^{-7}	3.6×10^{-7}	1.5×10^{-7}	0.051
30 Years	1.3×10^{-6}	1.8×10^{-6}	2.1×10^{-6}	2.1×10^{-6}	1.6×10^{-6}	1.0×10^{-6}	4.8×10^{-7}	1.5×10^{-7}	0.14
Life	3.6×10^{-6}	3.5×10^{-6}	3.0×10^{-6}	2.5×10^{-6}	1.7×10^{-6}	1.0×10^{-6}	4.8×10^{-7}	1.5×10^{-7}	0.21

[a] Radon to radon daughter ratio Rn/RaA/RaB/RaC equal to 1/0.9/0.7/0.7; unattached $\frac{RaA}{Rn}$ equal to 0.07.

[b] For a population with age characteristics equal to that in the whole United States in 1975.

$$R = (1 + 0.0031 \times \text{WLM})\ (1 + 0.00051 \times \text{PACKS}) \qquad (10\text{-}6)$$

where WLM and PACKS were total exposure to radon daughters and cigarettes from start of exposure to ten years before current age. Using this relative risk factor for radon daughters (1.0031 per WLM), 1975 U.S. population age specific lung cancer rates, a five year minimum latent interval and correcting for competing risks of death, the lifetime risk of lung cancer for an environmentally exposed population using the same exposure patterns as those in Table 10.2 were calculated. Lifetime lung cancer risk in the U.S. population for exposure to 1 WLM/year from birth using this relative risk model is essentially the same as the value given in Table 10.2 (0.009 vs. 0.0091). The lifetime risk for a shorter exposure duration is constant up to age 50 regardless of age at first exposure and then decreases somewhat more slowly than the modified absolute risk model. Thus, the relative risk model does not accommodate the higher risk for older ages with shorter term versus longer term exposure seen in the Czech miners as successfully as the modified absolute risk model. However, for the purpose of risk projection for environmental exposure of the population, the numerical values for lifetime risk are similar.

10.5 The Effect of Smoking upon Radon Daughter Induced Lung Cancer

Tables 10.1 to 10.3 have been developed without regard to differences in lifetime lung cancer risk between smokers and nonsmokers. Axelson and Sundell (1978) report that, for a small number of lung cancer cases developing in zinc-lead miners in Sweden, the lifetime risk for nonsmokers appears to be higher than for smokers. The average time of appearance of the tumors in smokers, however, was nine years earlier than in nonsmokers. A possible protective effect of smoking is supported by experimental data obtained with beagle dogs (7 vs. 1 tumor in 19 dogs) exposed to radon daughters and uranium ore dust with and without cigarette smoke (see Section 9). Axelson and Sundell (1968) tentatively ascribed the protective effect to a thickened mucus barrier in the airways, but suggested that, once initiated, the promotional effect of tobacco smoke causes tumors to appear faster.

The effect of smoking upon lung cancer appears sensitive to follow-up time. Based upon the study of Radford (1981) only small differences appear to exist with long-term follow up. The details of this Swedish iron miner study are reported in Section 8 (Section 8.3.3). He found the absolute risk to be 20×10^{-6} and 16×10^{-6} per person per year per WLM for smokers and nonsmokers respectively.

The model developed in this report can be modified to accommodate smokers and nonsmokers by selecting different values for tumor growth time rather than a single value of five years, and different coefficients rather than the single value of 10×10^{-6} per year per WLM. Not enough data are presently available to model U.S. smokers versus nonsmokers with any certainty. Using the relative risk model of Whittemore and McMillan (1983) [Eq. (10-6)] and U.S. age specific lung cancer rates for a population of nonsmokers, a lifetime lung cancer risk one-fourth of that in Table 10.2 would be projected for nonsmokers. The study of Radford (1981), however, suggests that there is possibly little difference in radiation-induced cancer between smokers and nonsmokers.

10.6 Calculated Lung Cancer Risk in Typical Environmental Situations

From Table 10.3 it can be seen that a lifetime exposure to an average of, say, 500 pCi $^{222}Rn/m^3$ would result in a lung cancer risk of 0.0018 or about 0.2 percent. Enstrom and Godley (1980) and Garfinkel (1980) have reported the annual age adjusted lung cancer rates in the United States for nonsmokers, that is, spontaneous lung cancer. From their data, a lifetime risk (age 40–85) of lung cancer for men and women may be calculated. These risks are 0.01 and 0.005, and 0.006 and 0.004, respectively, for the two studies. Hirayama (1981), in a study conducted in Japan, showed that nonsmoking wives of smoking husbands have a higher risk of lung cancer by a factor of two than wives of nonsmoking husbands. Garfinkel (1981) did not find this effect of passive smoking in U.S. nonsmoking data. The lifetime risks for nonsmokers reported by Enstrom and Godley (1980) and Garfinkel (1980) may or may not be somewhat elevated because of passive smoking. Their values, however, indicate that a typical environmental level of radon daughters could account for about one-fifth of the spontaneous lung cancer.

10.7 Summary

The model developed here is intended to utilize the lung cancer information obtained in epidemiological studies of underground miners at high levels of radon daughter exposure for extrapolation to risks at environmental levels. The criterion for the model is that it should fit the existing underground mining lung cancer data well. A model which expresses lung cancer risk uniformly with time after exposure (with the restriction that tumors do not occur either before a five-year latent interval or before age 40) and corrected from year of exposure by an exponential factor (20 year half-time) and an appropriate life table value to account for competing risks of death, satisfies this criterion. Lifetime lung cancer risks per rad/year, per WLM/year, and per pCi $^{222}Rn/m^3$ are then readily tabulated for different exposure intervals and are given in Tables 10.1, 10.2 and 10.3.

Lung cancer risk from environmental radon daughter exposure may be calculated from any of the tables. Lung cancer risk for underground mining may be calculated from Table 10.1 using the conversion factor specific to mining of 0.5 rad per WLM.

11. Evaluation of Occupational and Environmental Exposures

11.1 Introduction

The preceding chapters have reported on the sources of ^{222}Rn and its daughters occurring naturally and occupationally and on the alpha dose delivered to target cells involved in lung cancer induction. Experimental studies using animals have been reported and these have given insight into the lifetime risk of lung cancer induction by radon daughters at relatively high concentrations. Currently, studies at the CEN in France and at the Battelle Pacific Northwest Laboratory are being designed or conducted to investigate lifetime lung cancer induction in rats in the very low exposure region (<20 WLM).

The epidemiological studies of humans exposed to short-lived ^{222}Rn daughters in underground mines are still in progress even after 20 or more years. These will take perhaps another 20 years for essentially complete follow-up. The most complete existing study is that of Swedish iron miners (Radford and Renard, 1981) where 540 of 1290 miners followed are deceased and the remaining miners are in their 70's. So far 50 lung cancers have been observed with about 10 expected. It is unlikely that the total number of attributable lung cancers will double according to the present rate of appearance in the remaining 750 miners.

Although there is uncertainty in predicting lifetime risk per unit exposure, it is possible to estimate average values based upon present human and animal data.

11.2 Lifetime Risk for Occupational and Environmental Exposures

The human epidemiological studies with longest follow up can be used to calculate the lifetime risk of attributable lung cancer. The U.S.

uranium miner cohort of 3362 miners with 150 attributable lung cancers through December 1977, had an average cumulative exposure of 800 WLM and a median cumulative exposure of 400 WLM (Waxweiler, 1981). The Czechoslovakian reports show 202 attributable lung cancer deaths in 2500 miners through December 1975 (Kunz, 1979). The average exposure for all groups is not reported but can be estimated from the data as 300–400 WLM. The Swedish iron miner cohort has 40 attributable lung cancers out of 1290 miners followed through December 1976 and an estimated cumulative exposure of ~ 100 WLM.

Recent lifetime studies with rats have been reported by Chameaud *et al.* (1981) including cumulative exposures as low as 65 WLM.

The lifetime risks at low cumulative exposures from all of these studies and unpublished data supplied from the Battelle experiments with rats and dogs are plotted per unit exposure in Fig. 11.1. Both human and animal data suggest that the lifetime lung cancer risk is between 1 and 4 × 10^{-4} per WLM. Although the animal data are not corrected for life span shortening, the projected risk at low cumulative exposures is not expected to be altered appreciably.

The cumulative exposures in WLM for humans and animals are comparable on a dosimetric basis when the different sites for tumors are considered, that is, the first few branching airways in humans and the most peripheral airways in the hamster and rat.

One of the goals of this report was to evaluate the risks from present exposure to radon daughters, both for occupational conditions and for

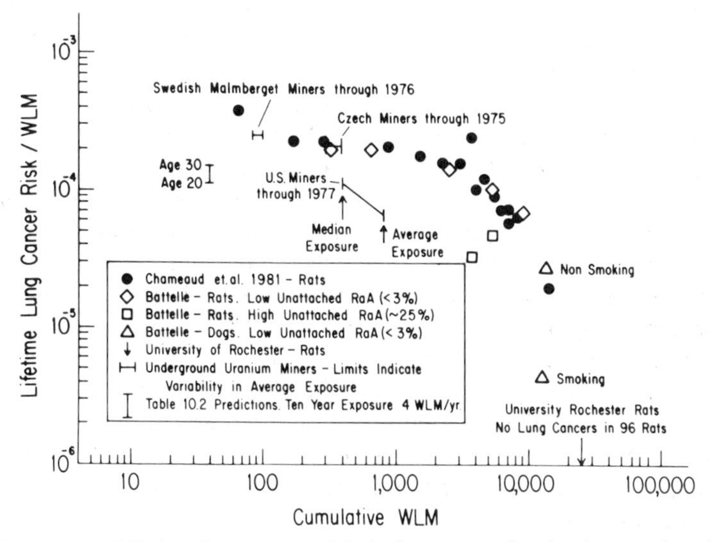

Fig. 11.1. Lifetime lung cancer risk in humans and animals as a function of cumulative radon daughter exposures.

the general public. To accomplish this, a comparison is also made with the current recommended limits for such exposures.

11.2.1 *Occupational Exposure*

All of the epidemiological studies on miners have related the observed attributable lung cancer production to the estimated cumulative exposure of the individual expressed in working level months. The recommended limit for occupational exposure in the United States is 4 WLM/year. Considering that lung cancer has definitely been produced at cumulative levels of 100 WLM, exposure at the present limit for a working lifetime of 50 years as used in radiation protection does not provide a factor of safety.

Waxweiler (1981) and Beverly (1981) have indicated that U.S. uranium miners have had average underground mining experience of less than 10 years. Based upon the present U.S. occupational standard of 4 WLM/year this yields a projected total cumulative exposure of less than 40 WLM.

The risk for the total cumulative exposure of 40 WLM can be estimated from the data for environmental exposures in Table 10.2. For 40 WLM accumulated in ten years the lifetime risk is 0.6 to 0.8 percent (0.006 to 0.008) depending upon whether age at first exposure is 20 or 30. This must be converted by the dose factor ratio for occupational versus environmental conditions of 0.5/0.7 as described in Section 10 to obtain a lifetime risk of from (0.6–0.8) (0.5/0.7) = 0.4 to 0.6 percent (0.004–0.006). The predicted risk (Table 10.2) for the U.S. miners for 40 WLM is also indicated on Fig. 11.1.

ICRP (1977) has reviewed the risk of death due to all causes in safe industries and found this risk to average 10^{-4} per year. This corresponds to a lifetime risk of death of about 0.5 percent for 50 years of employment. If the limit of 4 WLM/year is observed, then the average exposure should be considerably less but the recommended limit can be considered to be consistent with the 0.5 percent risk for ten years of exposure. Actual annual average exposures in U.S. uranium mines are presently reported to be between 1 and 2 WLM/year.

From Table 10.2, it may be seen that exposure at 1 WLM/year for 50 years would give a lifetime occupational lung cancer risk of (0.77%) (0.5/0.7) = 0.55%. This is comparable to the limits accepted by ICRP (1977) of 10^{-4} for the risk of death in "safe industries." Mining is certainly not a safe industry and the risk level used for comparison would be considerably higher.

External gamma-ray exposure in U.S. mines is not expected to contribute significantly to lifetime lung cancer risk (see Section 4.1).

ICRP (1977b) has adopted a lifetime total cancer risk factor of 100×10^{-6} per rem and a lung cancer risk of 20×10^{-6} per rem for external gamma ray exposure. Based on the values given in Section 4.1, a cumulative gamma-ray exposure for underground uranium miners should not reach a value of 50 rem. Therefore, a total cancer risk of 0.5 percent and a lifetime lung cancer risk of 0.1 percent from gamma ray exposure in mines is thought to represent an upper bound. Gamma ray dose measurements in individual mines should be performed, however, to substantiate the soundness of the above estimate.

11.2.2 *Exposure of the Public*

The average exposure of the public to radon daughter products in the United States has been estimated as 0.2 WLM/year in several studies. This exposure would give a lifetime risk of death from lung cancer of 1800 per million exposed (Table 10.2). This is a small fraction of the present incidence and is about one-fifth of the incidence in nonsmokers (see Section 10.6).

There has been concern expressed over the possible increase in public exposure in energy-efficient (low ventilation) homes or living in homes with elevated levels of radon daughters. There are insufficient data to evaluate these exposure increases, but it is possible to estimate the added risk from an arbitrary increase in the average exposure of the public.

Radiation protection recommendations in the past have largely been concerned with exposure of the worker. One limit for the population that has frequently appeared is that the population as a whole should not be exposed to incremental levels of radiation, above background, greater than 1/30th of the occupational limit. In the case of radon daughter exposure the arbitrary increase might be taken as (4) (1/30) or 0.13 WLM/year. A lifetime exposure at this increase in level would be estimated to result in an increase in lung cancer deaths of about 1200 per million, or over half of the estimated incidence due to average background.

This arbitrary increase of 0.13 WLM/year can quite possibly occur with significant change in home ventilation or with other changes in living conditions. Thus the case for radon daughter exposure is different from the usually considered population exposure to external gamma-ray radiation.

Recommendations on the exposure to radon daughters for individuals in the population are given in a separate NCRP report (NCRP, 1984).

APPENDIX A

Glossary

absorbed dose: The energy from ionizing radiation absorbed per unit mass is called the absorbed dose. The special unit of absorbed dose is the rad and is equal numerically to 100 erg/gram (10^{-2} Joule/kg).

aerodynamic diameter: The diameter of a unit density particle that has the same settling velocity as the particle described.

alpha particle: The nucleus of a helium atom which is ejected from some radionuclides during radioactive decay.

basal cell: The target cell at risk in lung cancer production whose minimum depth is of the order of 35 μm below the surface of the mucus layer or 22 μm below the surface of the bronchial epithelium. The basal cells are the dividing stem cells which produce replacement cells for those lost normally from the bronchial epithelium.

bronchial epithelium: The surface layer of cells lining of the conducting airways. The thickness decreases with bronchial generation from about 80 μm in the trachea to 15 μm in the finest airways.

bronchial generation: The branches of airways from the trachea to the gas exchange region of the lung. The branches are numbered sequentially from the trachea (generation 0).

cilia: The hairlike structures on some bronchial surface cells which propel the mucus and deposited material upward to the pharynx.

Curie (Ci): The special unit of radioactivity. One curie equals 3.7×10^{10} nuclear transformations per second.

diffusion: Movement by Brownian or random motion.

dose conversion factor: The absorbed dose in rad per working level month.

dosimetry: The measurement or calculation of the energy absorbed by matter.

epidemiology: The study of health and illness in human populations.

emanation rate: In this report, the release rate of radon from a surface in pCi/cm^2·sec.

environmental exposure: Exposure to radiation in nonoccupational situations.

epithelium: The surface cells that line the airways.

equilibrium: Equilibrium exists when the activity of all the short-lived radon daughters is equal to the parent radon activity. This is rarely achieved and the daughter activities are usually less than the radon activity.

lognormal distribution: The distribution of a random variable whose logarithm obeys the normal or Gaussian law of probability.

MeV: A unit of energy. The energy acquired by an electron accelerated through a potential difference of one million volts.

occupational exposure: Exposure of the worker during a period of work.

rad: The special unit of radiation absorbed dose. One rad equals the absorption of 100 ergs/gram of absorbing material (10^{-2} Joule/kg).

radon: In this report referring to the radioactive noble gas ^{222}Rn.

radon daughters: In this report, the short-lived radionuclides formed as a result of decay of ^{222}Rn. They consist of ^{218}Po (RaA), ^{214}Pb (RaB), ^{214}Bi (RaC) and ^{214}Po (RaC′). Their effective combined half-life is about 30 minutes.

relaxation length: The depth of soil which delivers a surface flux of radon that is 1/e of that delivered by soil at the surface.

risk coefficient: In this report, the attributable mortality rate of lung cancer per unit exposure following a suitable latent interval. In this report expressed in lung cancer deaths per year per WLM.

respiratory system: The organ system concerned with the exchange of oxygen and carbon dioxide. The respiratory system consists of the air passages through the nose, mouth, pharynx, the conducting airways (trachea, bronchi, bronchioles) and the gas exchange surfaces (alveoli) in the pulmonary parenchyma.

stochastic: Where the probability of an effect occurring rather than its severity is regarded as a function of dose without threshold.

tidal air volume: Volume of air inhaled per breath.

tracheobronchial tree: The portion of the respiratory system which conducts air to the alveoli.

unattached RaA (^{218}Po) or RaA*: The fraction of the equilibrium amount of RaA which is not attached to ambient aerosol.

working level (WL): Any combination of short-lived radon daughters in one liter of air that will result in the emission of 1.3×10^5 MeV of potential alpha energy.

working level month (WLM): The cumulative exposure equivalent to exposure to one working level for a working month (170 hours).

References

AEC (1951). Atomic Energy Commission, "Symposium, The Biophysics of Radium and Radon," Original transcript of The Radium Meeting held at the New York operations office, Atomic Energy Commission, 13–14 June 1951 (U.S. Atomic Energy Commission, Washington, D.C.).

AEC (1960). Atomic Energy Commission, *Code of Federal Regulations,* 10 CFR Part 20 (Government Printing Office, Washington).

ALTSHULER, B., NELSON, N. AND KUSCHNER, M. (1964). "Estimation of lung tissue dose from the inhalation of radon and daughters," Health Phys. **10,** 1137.

ASA (1952). American Standards Association (now American National Standards Institute) ASA-Z37, May 1952, Cited in Joint Committee on Atomic Energy, Subcommittee on Research, Development, and Radiation, Congress of the United States, Ninetieth Congress, May–August 1967: Hearings on Radiation Exposure of Uranium Miners, Part 2, page 1217 (Government Printing Office, Washington, D.C.).

ASA (1960). American Standards Association (now American National Standards Institute), "Radiation protection in uranium mines and mills (concentrators)." ASA-N7.1-1960 (American National Standards Institute, New York, New York).

ARCHER, V. E., WAGONER, J. K. AND LUNDIN, F. E. (1973). "Lung cancer among uranium miners in the United States," Health Phys. **25,** 351.

ARCHER, V. E., SACCOMANO, G. AND JONES, J. H. (1974). "Frequency of different histologic types of bronchogenic carcinoma as related to radiation exposure," Cancer **34,** 2056.

ARCHER, V. E., GILLAM, J. D. AND WAGONER, J. K. (1976). "Respiratory disease mortality among uranium miners," Ann. N.Y. Acad. Sci. **271,** 280.

ARCHER, V. E., GILLAM, J. D. AND JAMES, L. A. (1978). "Radiation and smoking relationships to lung cancer in uranium miners." page 1689 in *Proceedings of the Third International Symposium on Detection and Prevention of Cancer*, Nieburgs, H. E., Ed. (Marcel Dekker, New York).

ARCHER, V. E., RADFORD, E. P., AND AXELSON, O. (1979). "Radon daughter cancer in man: Factors in exposure response relationship," page 324 in *Conference Workshop on Lung Cancer Epidemiology and Industrial Applications of Sputum Cytology* (Colorado School of Mines Press, Golden, Colorado).

ARCHER, V. E. (1981). "Health concerns in uranium mining and milling," J. Occup. Medicine **23,** 502.

ASMUNDSSON, T. AND KILBURN, K. H. (1970). "Mucociliary clearance rates at various levels in dog lungs," Am. Rev. Respir. Disease **102,** 388.

AXELSON, O. (1979). "Calculations based on the Hammar Parish population as including most of the Zincgruvan miners," Unpublished data.

AXELSON, O. AND SUNDELL, L. (1978). "Mining lung cancer and smoking," Scand. J. Work Environ. and Health **4,** 46.

AXELSON, O. AND EDLING, C. (1980). "Health hazards from radon daughters in Sweden," in *Health Implication of New Energy Technologies*, Proceedings of an environmental health conference, April 4–7, 1979, Washington, D. C. (Ann Arbor Science Publishers, Ann Arbor, Michigan).

BALE, W. F. (1951). *Hazards Associated with Radon and Thoron*, Memo, Div. Biol. and Med., Atomic Energy Commission, Washington, D. C. (March 14, 1951).

BALE, W. F. AND SHAPIRO, J. V. (1956). "Radiation dosage to lungs from radon and its daughter products," page 233 in *Proceedings 2nd U.N. International Conference on Peaceful Uses of Atomic Energy*, Vol. 13 (United Nations Publications, New York, New York).

BARNARD, W. C. (1926). "The nature of the oat-celled sarcoma of the mediastinum," J. Path. Bacteriol. **29,** 241.

BARNETT, B. AND MILLER, C. E. (1966). "Flow induced by biological mucociliary systems," Ann. N. Y. Acad. Sci. **130,** 891.

BARRETTO, P. M. C., CLARK, R. B. AND ADAMS, J. A. S. (1972). "Physical characteristics of radon-222 emanation from rocks, soils, and minerals: Its relation to temperature and alpha dose," page 731 in *The Natural Radiation Environment II*, U.S. Energy Research and Development Administration Report CONF 720805 (National Technical Information Service, Springfield, Virginia).

BARTON, C. J., MOORE, R. E. AND ROHWER, P. S. (1973). *Contribution of Radon in Natural Gas to the Natural Radioactivity Dose in Homes*, USAEC Report ORNL-TM-4154 (National Technical Information Service, Springfield Virginia).

BECK, H. L. (1974). "Gamma radiation from radon daughters in the atmosphere," J. Geophys. Res. **79,** 2215.

BECK, H. L., GOGOLAK, C. V., MILLER, K. E. AND LOWDER, W. M. (1978). "Pertubations on the natural environment due to the utilization of coal as an energy source," (presented at Natural Radiation Environment III, Houston, Texas, April 1978).

BEEBE, G. W., KATO, H., AND LAND, C. E. (1978). "Studies of the mortality of A-bomb survivors. 6. Mortality and radiation dose, 1950–1974," Radiat. Res. **75,** 138.

BEHOUNEK, F. (1969). "History of the exposure of miners to radon," Health Phys. **19,** 56.

BENSCH, K. G., GORRIN, B., PARIENTE, R. AND SPENCER, H. (1968). "Oat-cell carcinoma of the lung," Cancer **22,** 1163.

BIROT, A. (1971). *Contribution a l'etude de la diffusion du radon et des aerosols dans la troposphere*, Thesis (Universite de Toulouse, France).

BLAKE, J. (1975). "On the movement of mucus in the lung," J. Biomech. **8,** 179.

BOCKENDAL, A. (1885). "Uber die Regeneration des Trachealepithels," Arch. Mikrosk Anat. **24,** 361.

BOUCHARD, C., CURIE, P. AND BALTHAZARD, V. (1904). "Action physiologique de l'emanation du radium," C.R. Acad. Sci. **138,** 1384.

BOYD, J. T., DOLL, R., FAULDS, J. S. AND LEIPER, J. (1970). "Cancer of the lung in iron or (hematite) miners," Br. J. Med. **27,** 29.

BREEZE, R. C. AND WHEELDON, E. B. (1977). "The cells of the pulmonary airways," Am. Rev. Respir. Dis. **116,** 705.

BXRPC (1943). British X-Ray and Radium Protection Committee, (Sixth Report), Cited in Joint Committee on Atomic Energy, Subcommittee on Research, Development, and Radiation, Congress of the United States, Ninetieth Congress, May–August 1967: Hearing on Radiation Exposure of Uranium Miners, Part 2, page 1217. (Government Printing Office, Washington, D.C.).

BXRPC (1948). British X-Ray and Radium Protection Committee, (Seventh Revised Report), Cited in Joint Committee on Atomic Energy, Subcommittee on Research, Development, and Radiation, Congress of the United States, Ninetieth Congress, May–August 1967: Hearings on Radiation Exposure of Uranium Miners, Part 2, page 1217. (Government Printing Office, Washington, D.C.).

CASTREN, O. (1980). "The Contribution of Bored Wells to Respiratory Radon Daughter Exposure in Finland," page 1364 in *Natural Radiation Environment III*, Gesell T. F. and Lowder, W. M., Eds. (National Technical Information Service, Springfield, Virginia).

CHAMBERLAIN, A. C. AND DYSON, E. D. (1956). "The dose to the trachea and bronchi from the decay products of radon and thoron," Br. J. Radiol. **29,** 317.

CHAMEAUD, J., PERRAUD, R., MASSE, R., NENOT, J. C. AND LAFUMA, J. (1976). "Lung cancer induced in rats by radon and its daughter nuclides at different concentrations," page 223 in *Biological and Environmental Effects of Low Level Radiation*, Vol. II, International Atomic Energy Agency Publication STI/PUB/409 (International Atomic Energy Agency, Vienna).

CHAMEAUD, J., PERRAUD, R., CHETIEN, J., MASSE, R. AND LAFUMA, J. (1978). "Experimental study of the combined effect of cigarette smoke and an active burden of radon-222," page 429 in *Late Biological Effects of Ionizing Radiation*, Vol. II, IAEA Publication STI/PUB/489 (International Atomic Energy Agency, Vienna).

CHAMEAUD, J., PERRAUD, R., MASSE, R. AND LAFUMA, J. (1981). "Contribution of Animal Experimentation to the Interpretation of Human Epidemiological Data," Proc. of International Conference on Radiation Hazards in Mining—Control, Measurement and Medical Aspects, Golden Colorado, Oct. 5–9, 1981 (Society of Mining Engineers, New York, New York).

CHAMEAUD, J., PERRAUD, R., MASSE, R. AND LAFUMA, J. (1981). "Contribution of Animal Experimentation to the Interpret of Human," *Epid. Data Proc. Int. Conf. on Rad. Haz.*

CHAN, T. L. (1977). *The Characterization of Particle Deposition in the Human*

Tracheobronchial Tree, Doctoral dissertation (Dept. of Environmental Health Sciences, New York University).

CHATEBAKIS, C. B. (1960). "The incidence of malignant disease in 1620 African Bantu gold miners," Arch. Environ. Health **1,** 17.

CHOVIL, A. (1981). "The Epidemiology of Primary Lung Cancer in Uranium Miners in Ontario," J. Occupational Med. **23,** 417.

CFR (1972). Code of Federal Regulations, Title 30—Part 57, *Mineral Resources*, revised January 1, 1972 (Government Printing Office, Washington, D.C.).

CLIFF, K. D. (1978). "Assessment of airborne radon daughter concentrations in dwellings in Great Britain," Phys. Med. Biol. **23,** 696.

COHEN, A. F. AND COHEN, B. L. (1980). "Tests of the linearity assumption in the dose-effect relationship for radiation-induced cancer," Health Phys. **38,** 53.

COHN, S. H., SKOW, R. K. AND GONG, J. K. (1953). "Radon inhalation studies in rats," Arch. Ind. Hyg. Occup. Med. **7,** 508.

COLES, D. G., RAGAINI, R. C. AND ONDOV, J. M. (1978). "Behavior of natural radionuclides in western coal-fired power plants," Environ. Sci. Tech. **12,** 442.

COO (1968). Colorado State University, *Radon Progeny Inhalation Study as Applicable to Uranium Mining Operations*, Colorado State University Report COO-1500-5 (National Technical Information Service, Springfield, Virginia).

COOPER, J. A., *ET AL.* (1973). *Characteristics of Attached Radon-222 Daughters under Both Laboratory and Field Conditions with Particular Emphasis upon Underground Uranium Mine Environments*, Report EN-SA-299 (Battelle Pacific Northwest Laboratory, Richland, Washington).

CROSS, F. T. (1979). "Exposure standards for uranium," Health Phys. **37,** 765.

CROSS, F. T., BLOOMSTER, C. H., HENDRICKSON, P. L., NELSON, I. C., HOOPER, B. L., MERRILL, J. A., AND STUART, B. O. (1974). *Evaluation of Methods for Setting Occupational Health Standards for Uranium Mines*, Appendix A. Research Report prepared for the National Institute for Occupational Safety and Health (Battelle Pacific Northwest Laboratory, Richland, Washington).

CROSS, F. T., PALMER, R. F., FILIPY, R. E., BUSCH, R. H. AND STUART, B. O. (1978). *Study of the Combined Effects of Smoking and Inhalation of Uranium Ore Dust. Radon Daughters and Diesel Oil Exhaust Fumes in Hamsters and Dogs,* Battelle Pacific Northwest Laboratory Final Report PNL-2744/UC-48 (National Technical Information Service, Springfield, Virginia).

CUTZ, E., CHAN, W., WONG, V. AND CONEN, P. E. (1975). "Ultrastructure and fluorescence histochemistry of endocrine (QPUD-Type) cells in tracheal mucosa of human and various animal species," Cell Tissue Res. **158,** 425.

DAVIES, C. N. (1946). "Filtration of droplets in the nose of the rabbit," Proc. R. Soc. London, Ser. B **133,** 282.

DAVIES, C. N. (1973). "Diffusion and sedimentation of aerosol particles from Poiseuille flow in pipes," J. Aerosol Sci. **4,** 317.

DESROSIERS, A. E. (1977). "Alpha particle dose to respiratory airway epithelium," Health Phys. **32,** 192.

DESROSIERS, A. E., KENNEDY, A. AND LITTLE, J. B. (1978). "Rn-222 daughter dosimetry in the Syrian Golden hamster lung," Health Phys. **35,** 607.

DEVILLIERS, A. J. AND WINDISH, J. P. (1964). "Lung cancer in a fluorspar mining community: I. Radiation, dust, and mortality experience," Br. J. Ind. Med. **21,** 94.

DIBNER, B. (1958). *Agricola on Metals, De Re Metallica* by Georgius Agricola, Basel, 1597, translated by Herbert C. Hoover (Burndy Library, Inc., Norwalk, Connecticut).

DRASCH, O. (1881). "Zur Fragee der Reegeneration des Tracheenlepithels mit Rucksicht auf die Karyokinase und die Bedentung der Becherzallen," Sitzurgsber. Acad. Wiss Wien Math Naturwiss Kl Abt **3:83,** 341.

DUGGAN, M. J. AND HOWELL, D. M. (1969a). "The measurement of the unattached fraction of airborne RaA," Health Phys. **17,** 423.

DUGGAN, M. J. AND HOWELL, D. M. (1969b). "Relationship between the unattached fraction of RaA and the concentration of condensation nuclei," Nature **224,** 1190.

DUGGAN, M. J., SOILLEUX, P. J., STRONG, J. C. AND HOWELL, D. M. (1970). "The exposure of United Kingdom miners to radon," Br. J. Ind. Med. **27,** 106.

DUNCAN, D. L., BOYSSEN, G. A., GROSSMAN, L. AND FRANZ, G. A. III. (1977). *Outdoor Radon Study (1974–1975). An Evaluation of Ambient Radon-222 Concentration in Grand Junction, Colorado, Final Report*, Environmental Protection Agency Report ORP/LV-77-1 (National Technical Information Service, Springfield, Virginia).

EADIE, G. G., FORT, C. W. AND BEARD, M. L. (1979). *"Airborne Radioactivity Measurements in the Vicinity of the Jackpile Open Pit Uranium Mine, New Mexico,* Environmental Protection Agency Report ORP/LV-79-2 (National Technical Information Service, Springfield, Virginia).

EADIE, G. G., KAUFMANN, R. F., MARKLEY, D. J. AND WILLIAMS, R. (1976). *Report of Ambient Outdoor Radon and Indoor Radon Progeny Concentration During 1975 at Selected Locations in the Grants Mineral Belt, New Mexico, Final Technical Note*, Environmental Protection Agency Report ORP/LV-76-4 (National Technical Information Service, Springfield, Virginia).

EDITORIAL (1970). "Lung cancer in hematite miners," The Lancet **7676,** 785.

EDLING, C. (1982). "Lung cancer and smoking in a group of iron ore miners," *American J. Ind. Medicine* **3,** 191.

ENSTRON, J. E. AND GADLEY, F. H. (1980). "Cancer mortality among a representative sample of non-smokers in the United States during 1966–68." J. Nat. Cancer. Inst. **65,** 1175.

EPA Office of Radiation Programs (1973). *Environmental Analysis of the Uranium Fuel Cycle*, Environmental Protection Agency Report EPA-520/9-73-B (U.S. Environmental Protection Agency, Washington, D.C.).

EVANS, R. D. (1950). "Quantitative aspects of radiation carcinogenesis in humans," Acta Union Int. Contra Cancrum **6,** 1229.

EVANS, R. D. AND GOODMAN, C. (1940). "Determination of the thoron content of air and its bearing on lung cancer hazards in industry," J. Ind. Hyg. Toxicol. **22,** 89.

EVANS, R. D., HARLEY, J. H., JACOBI, W., MCLEAN, A. S., MILLS, W. A. AND STEWART, C. G. (1981). "Estimate of Risk from Environmental Exposure to Radon-222 and its Decay Products," Nature **290,** 98.

FAULDS, J. S. AND STEWART, M. J. (1956). "Carcinoma of the lungs in hematite miners," J. Pathol. Bacteriol. **72,** 353.

FEYRTER, F. (1954). "Zur Pathologie des Argyrophilen Helle-Zellen-Organs im Bronchialbaum des Menchen," Virchows Arch. Pathol. Anat. Physiol. **325,** 723.

FISENNE, I. M. (1980). "Radon-222 measurements at Chester," page 73 in *Environmental Measurements Laboratory Regional Baseline Station, Chester, New Jersey,* Department of Energy Report EML-383. (National Technical Information Service, Springfield, Virginia).

FISENNE, I. M. AND HARLEY, N. H. (1974). *Lung Dose Estimates from Natural Radioactivity Measured in Urban Air,* HASL-TM74-7 (unpublished).

FITZGERALD, J. E., JR. AND SENSINTAFFAR, E. L (1977). "Radiation exposure from construction materials using byproduct phosphogypsum," in *Symposium on Public Health Aspects of Radioactivity in Consumer Products,* Department of Energy Report CONF 770208 (National Technical Information Service, Springfield, Virginia).

FOX, A. J. AND COLLIER, P. F. (1976). "Low mortality Rates in Industrial cohort studies due to selection for work and Survival in the industry," Br. Prev. Soc. Med. **30,** 225.

FOX, A. J., GOLDBLATT, P. AND KINLEN, L. J. (1981). "A Study of the Mortality of Cornish Tin Miners," Brit. J. Ind. Med. **38,** 378.

FR (1967a). *Federal Register,* Vol. 32, No. 113.

FR (1967b). *Federal Register,* Vol. 32, No. 147.

FR (1969a). *Federal Register,* Vol. 34, No. 10.

FR (1969b). *Federal Register,* Vol. 34, No. 11.

FR (1971). *Federal Register,* Vol. 36, No. 101.

FR (1971). *Federal Register,* Vol. 34, No. 576.

FR (1971). *Federal Register,* Vol. 36, No. 9480.

FRC (1967). Federal Radiation Council, *Guidance for the Control of Radiation Hazards in Uranium Mining,* Report No. 8, Revised (U.S. Government Printing Office, Washington, D.C.).

GABUNIYA, G. S. (1969). "Background radiation in Georgian mines," Gig. Sanit. **34,** 236, as translated by Israel Program for Scientific Translations (1970) (National Technical Information Service, Springfield, Virginia).

GARFINKEL, L. (1980). "Cancer mortality in non-smokers: Prospective study by the American Cancer Society," J. Nat. Cancer Inst. **65,** 1169.

GARFINKEL, L. (1981). "Time trends in lung cancer mortality among non-smokers and a note on passive smoking," J. Nat. Cancer Inst. **66,** 1061.

GASTINEAU, R. M. (1969). *Investigation of the Thickness of Bronchial Epithelium*, Doctoral Thesis (University of North Carolina, Chapel Hill, North Carolina).

GASTINEAU, R. M., WALSH, P. J. AND UNDERWOOD, N. (1972). "Thickness of bronchial epithelium with relation to exposure to radon," Health Phys. **23,** 857.

GEORGE, A. C. (1972). "Measurements of the uncombined fraction of radon daughters with wire screens," Health Phys. **23,** 390.

GEORGE, A. C. (1975). "Indoor and outdoor measurements of natural radon and radon daughter decay products in New York City air," page 741 in *The Natural Radiation Environment II*, Vol. II, U.S. Energy Research and Development Administration Report CONF 720805 (National Technical Information Service, Springfield, Virginia).

GEORGE, A. C. AND HINCHLIFFE, L. (1972). "Measurements of uncombined radon daughters in uranium mines," Health Phys. **23,** 791.

GEORGE, A. C. AND BRESLIN, A. J. (1980). "The distribution of ambient radon and radon daughters in residential building in the New Jersey-New York area," in *The Natural Radiation Environment III*. Gesell, T. F. and Lowder, W. M., Eds. (Technical Information Center/U.S. Department of Energy).

GEORGE, A. C., HINCHLIFFE, L., EPPS, R. AND SHEPICH, T. J. (1970). *Respiratory Tract Deposition of Radon Daughters in Humans Exposed in a Uranium Mine (Phase II)*. Atomic Energy Commission, Health and Safety Laboratory Report TM70-7 (U.S. Atomic Energy Commission, Washington, D.C.).

GEORGE, A. C., HINCHLIFFE, L. AND SLADOWSKI, R. (1975). "Size distribution of radon daughter particles in uranium mine atmospheres," Am. Ind. Hyg. Assn. J. **36,** 4884.

GESELL, T. F. AND PRICHARD, H. M. (1980). "Contribution of radon in tap water to indoor radon concentrations," in *Natural Radiation Environment III*, Houston, Texas, April 1978. Gesell, T. F. and Lowder, W. M., Eds. (Technical Information Center/US Dept. of Energy).

GOLDZIECHER (1918). "Uber baselzellen wucherungea der bronchial schliemeit," Zentralblatt fur Allgemein Path. und Pathol. Anat. **29,** 506.

GOODMAN, R. E., YERGIN, B. M., LANDA, J. F., GOLINVAUX, M. H. AND SACKER, M. A. (1978). "Relationship of smoking history and pulmonary function tests to tracheal mucus velocity in non-smokers, young smokers, exsmokers, and patients with chronic bronchitis," Am. Rev. Respir. Dis. **117,** 205.

GORMLEY, P. C. AND KENNEDY, M. (1949). "Diffusion from a stream through a cylindrical tube," Proc. R. Ir. Acad. Sect. **A 52,** 163.

GOTTLIEB, L. S. AND HUSON, L. A. (1982). "Lung Cancer among Navajo Uranium Miners," Chest **81,** 449.

GUIMOND, R. J. (1977). "Radiological aspects of fertilizer utilization," (presented at the Symposium on Public Health Aspects of Radioactivity in Consumer Products, Atlanta, Georgia, February 1977).

HAMRICK, P. E. AND WALSH, P. J. (1974). "Environmental radiation and the lung," Env. Health Perspectives **9,** 33.

HAQUE, A. K. M. M. (1967a). "Energy expended by alpha particles in lung tissue," Br. J. Appl. Phys. **17,** 905.

HAQUE, A. K. M. M. (1967b). "Energy expended by alpha particles in lung tissue. III. A computer method of calculation," Br. J. Appl. Phys. **18,** 657.

HAQUE, A. K. M. M. AND COLLINSON, A. J. L. (1967). "Radiation dose to the respiratory system due to radon and its daughter products," Health Phys. **13,** 431.

HARLEY, J. H. (1952). *Sampling and Measurement of Airborne Daughter Products of Radon,* Doctoral Dissertation (Rensselaer Polytechnic Institute).

HARLEY, J. H. (1953). "Sampling and measurement of air-borne daughter products of radon," Nucleonics **11,** 12.

HARLEY, J. H. (1973). "Environmental radon," page 109 in *Noble Gases,* Stanley, R. E. and Moghissi, A. A., Eds., National Environmental Research Center Report CONF-730915 (National Environmental Research Center, Las Vegas, Nevada).

HARLEY, J. H. (1978). *Regional Baseline Station, Chester, New Jersey,* Department of Energy Report EML-347 (National Technical Information Service, Springfield, Virginia).

HARLEY, J. H. (1979). *Regional Baseline Station, Chester, New Jersey,* Department of Energy Report EML-383 (National Technical Information Service, Springfield, Virginia).

HARLEY, N. H. (1980). "Comments on the proposed ICRP lung model as applied to occupational limits for short lived radon daughters: A comparison with epidemiologic and dosimetry models," Presented at the Berlin Colloquium, October 7–9, 1980.

HARLEY, N. H. (1972). "Alpha absorption measurement applied to lung dose from radon daughters," Health Phys. **23,** 771.

HARLEY, N. H. AND PASTERNACK, B. S. (1981). "A model for predicting lung cancer risks induced by environmental levels of radon daughters," Health Phys. **40,** 307.

HARLEY, N. H. AND PASTERNACK, B. S. (1982). "Environmental radon daughter alpha dose factors in five-lobed human lung," Health Phys. **42,** 789.

HARRIS, S. J. (1954). "Radon exposures in various mines" (Paper presented to the Ind. Health Conf., Chicago, Illinois April 28, 1954).

HASL (1960). *Experimental Environmental Study of AEC Leased Uranium Mines,* USAEC REPORT HASL-91 (Health and Safety Laboratory-now Environmental Measurements Laboratory, New York).

HATTORI, S., MATSUDA, M., TATEISHI, R., NISHIHARA, H. AND HARAI, T. (1972). "Oat cell carcinoma of the lung," Cancer **30,** 1014.

HESS, C. T., CASPARIUS, R. E., NORTON, S. A. AND BRUTSAERT, W. F. (1978). "The investigation of natural levels of ^{222}Rn in groundwater in Maine for assessment of related health effects," in *The Natural Radiation Environment III,* Gesell, T. F. and Lowder, W. M., Eds. (Technical Information Center/ Department of Energy).

HEWITT, D. (1976). "Radiogenic lung cancer in Ontario uranium miners 1955–1974," pages 78 and 319 in *Report of the Royal Commission on the Health and Safety of Workers in Mines,* Ham, J. A., Ed. (Ministry of the Attorney

General, Province of Ontario, Toronto, Canada).

HEWITT, D. (1979). "Biostatistical studies on Canadian uranium miners," pages 264 and 398 in *Conference/Workshop on Lung Cancer Epidemiology and Industrial Applications of Sputum Cytology* (Colorado School of Mines Press, Golden, Colorado).

HIRAYAMA, T. (1981). "Nonsmoking wives of heavy smokers have a higher risk of lung cancer: A study from Japan," Br. Med. J. **282,** 183.

Ho, W. L., Hopke, P. K. and Stukel, J. J. (1982). "The attachment of RaA (Po-218) to monodisperse aerosol," J. Aerosol Sci. **16,** 825.

HOLADAY, D. A. (1952). *Investigation of Radon and Radon Decay Products in the Uranium Mines of the Colorado Plateau* (Fed. Sec. Agency, P. H. S., Div. Occup. Med., undated work begun in 1952).

HOLADAY, D. A. (1955). "Digest of the proceedings of the seven-state conference on health hazards in uranium mining," Arch. Industr. Health **12,** 465.

HOLADAY, D. A. (1964). "Origin, history, and development of the uranium study," Memo prepared from material in P. W. Jacoe and D. A. Holaday files.

HOLADAY, D. A. (1969). "History of the exposure of miners to radon," Health Phys. **16,** 547.

HOLADAY, D., DAVID, W. AND DOYLE, E. N. (1952). *An Interim Report Health Study of the Uranium Mines and Mills* (Fed. Sec. Agency, P. H. S., Div. Occup. Health and the Colorado State Dept. of Health, May 1952).

HOLADAY, D. A., RUSHING, D. E., COLEMAN, R. D., WOOLRICH, P. F., KUSNETS, H. L. AND BALE, W. F. (1967). *Control of Radon and Daughters in Uranium Mines and Calculations on Biological Effects*, U.S. Dept. of Health, Education and Welfare, Public Health Service, Publ. No. 494 (Government Printing Office, Washington, D.C.).

HOLLEMAN, D. F. (1968). *Radiation Dosimetry for the Respiratory Tract of Uranium Miners*, Atomic Energy Commission Report COO-1500-12 (National Technical Information Service, Springfield, Virginia).

HOLMA, B. (1969). "Scanning electron microscope observation of particles deposited in the lung," Arch. Env. Health **18,** 330.

HORACEK, J. (1969). "Der Joachimstaler lungenkrebs nach dem zweiten weltkrieg (bericht uber 55 falle)," Krebsforsch **72,** 52.

HORACEK, J., PLACEK, V. AND SEVC, J. (1977). "Histologic types of bronchogenic cancer in relation to different conditions of radiation exposure," Cancer **40,** 832.

HORSFIELD, K., DART, G., OLSON, D. E., FILLEY, G. F. AND CUMMING, G. (1971). "Models of the human bronchial tree," J. Appl. Physiol. **31,** 207.

HORTON, T. R. (1977). "Estimation of Rn-222 daughter doses from large-area sources," Trans. Am. Nucl. Soc. **27,** 132.

HUEPER, W. (1942). *Occupational Tumors and Allied Diseases* (Charles C. Thomas, Springfield, Illinois).

HULTQVIST, B. (1956). "Studies on naturally occurring ionizing radiations," Kgl. Svenska Vetenskapsakad. Handl. Ser. 4, Band 6, No. **3,** Stockholm.

IAEA (1973). International Atomic Energy Agency, *Inhalation Risks from*

Radioactive Contaminants, Technical Report Series, No. 142 (International Atomic Energy Agency, Vienna).

ICRP (1955). International Commission on Radiological Protection, *Recommendations of the International Commission on Radiological Protection Supplement No. 6*, British Journal of Radiology, London.

ICRP (1959). International Commission on Radiological Protection, *Report of Committee II on Permissible Dose for Internal Radiation*, ICRP Publication 2 (Pergamon Press, New York).

ICRP (1966). International Commission on Radiological Protection, Task Group on Lung Dynamics, "Deposition and retention models for internal dosimetry of the human respiratory tract," Health Phys. **12,** 173.

ICRP (1974). International Commission on Radiological Protection, *Report of the Task Group on Reference Man*, ICRP Publication 23 (Pergamon Press, New York).

ICRP (1977a). International Commission on Radiological Protection, *Radiation Protection in Uranium and Other Mines.* ICRP Publication 24 (Pergamon Press, New York).

ICRP (1977b). International Commission on Radiological Protection, *Radiation Protection*, ICRP Publication 26 (Pergamon Press, New York).

ICRP (1978). International Commission on Radiological Protection, *Limits for Intake of Radionuclides by Workers.* ICRP Publication 30, Part I (Pergamon Press, New York).

ICRP (1981). International Commission on Radiological Protection, *Limits for Inhalation of Radon-222, Radon-220 and Their Short Lived Daughters.* ICRP Publication (Final Draft, Sept. 1980).

INGHAM, D. B. (1975). "Diffusion of aerosols from a stream flowing through a cylindrical tube," J. Aerosol Sci. **6,** 125.

ISRAEL, L. AND CHACHINIAN, A. P., (1976). *Lung Cancer, Natural History, Prognosis and Therapy* (Academic Press, New York).

JACKSON, M. L. (1940). *The Biological Effects of Inhaled Radon*, Master's Thesis (Massachusetts Institute of Technology).

JACOBI, W. (1964). "The dose to the human respiratory tract by inhalation of short-lived ^{222}Rn and ^{220}Rn decay products," Health Phys. **10,** 1163.

JACOBI, W. (1972). "Relations between the inhaled potential energy of ^{222}Rn and ^{220}Rn daughters and the absorbed energy in the bronchial and pulmonary region," Health Phys. **23,** 3.

JACOBI, W. (1973). "Relation between cumulative exposure to radon-daughters, lung dose and lung cancer risk," page 492 in *Noble Gases*, Stanley, R. E. and Moghissi, A. A., Eds., National Environmental Research Center Report CONF-730915 (National Environmental Research Center, Las Vegas, Nevada).

JACOBI, W. (1977). "Interpretation of measurements in uranium mines: dose evaluation and biomedical aspects," page 33 in *Proceedings of NEA Specialist Meeting. Personal Dosimetry and Area Monitoring Suitable for Radon and Daughter Products* (Nuclear Energy Agency, OECD, Paris).

JACOBI, W. and EISFELD, K. (1980). *Dose to Tissue and Effective Dose*

Equivalent by Inhalation of Radon-222 and Their Short-Lived Daughters, GSF Report S-626 (Gesellschaft fur Strahlen und Umweltforschung, Neuherberg, Federal Republic of Germany).

JAMES, A. C. (1976). "Bronchial deposition of free ions and submicron particles studies in excised lung," page 203 in *Inhaled Particles IV*, Walter, W. H., Ed. (Unwin Bros., Old Woking, Surrey, England).

JAMES, A. C. (1978). "Lung deposition of submicron aerosols calculated as a function of age and breathing rate," page 71 in *National Radiological Protection Board Annual Research and Development Report 1977*, NRPD/RD-02 (National Radiological Protection Board, Harwell, England).

JAMES, A. C., BRADFORD, G. E. and HOWELL, D. M. (1972). "Collection of unattached RaA atoms using a wire gauze," J. Aerosol Sci. **3,** 243.

JAMES, A. C., GREEHALGH, J. R. and SMITH, H. (1977). "Clearance of lead-212 ions from rabbit bronchial epithelium to blood," Phys. Med. Biol. **22,** 932.

JAMES, A. C., JACOBI, W. and STEINHAUSLER, F. (1981). "Respiratory Tract Dosimetry of Radon and Thoron Daughters: The State of the Art and Implications for Epidemiology and Radiobiology," page 42 in *Radiation Hazards in Mining: Control, Measurement and Medical Aspects.* Colorado School of Mines, Golden, Colorado, Oct. 1981 (Society of Mining Engineers, New York, New York).

JANSEN, H. and SCHULTZER, P. (1926). "Experimental investigations into the internal radium emanation therapy. I. Emanatorium experiments with rats," Acta Radiol. **6,** 631.

JCAE (1967). Joint Committee on Atomic Energy, Subcommittee of Research Development and Radiation, Congress of the United States, Ninetieth Congress, May–August 1967: *Hearings on Radiation Exposure of Uranium Miners* (cited on page 3315) (Government Printing Office, Washington, D.C.).

JONASSEN, N. and MCLAUGHLIN, J. P. (1978). "Exhalation of radon-222 from building materials and walls" (Presented at Natural Radiation Environment III, Houston, Texas, April, 1978).

JONES, T. D., KERR, G. D. and HWANG, J. M. L. (1978). "Chord operators for insult assessment to the radiosensitive cells of the tracheobronchial tree," page 28 in *Health Physics Division Annual Progress Report. Period Ending June 30, 1977*, Oak Ridge National Laboratory Report ORNL-53088 (National Technical Information Service, Springfield, Virginia).

JORGENSEN, H. S. (1973). "A study of mortality from lung cancer among miners in Kiruna, 1950–1970," Work Environment Health **10,** 126.

KIRICHENCO, V. N., KHACHIROV, D. ZH. G., DUBRUVIL, S. A. and YEKLYUCH, V. (1970). "Experimental study of the distribution of short-lived daughter products of radon in the respiratory tract," Gigirna I. Sanitaria **2,** 52 (also see Health Phys. **23,** 13).

KOTIN, P., COURINGTON, D. and FALK, H. L. (1966). "Pathogenesis of cancer in a ciliated mucus-secreting epithelium," Amer. Rev. Respir. Dis. **93,** Suppl. 115.

KRANER, H. W., SCHROEDER, G. L. and EVANS, R. D. (1964). "Measurements of the effects of atmospheric variables on radon-222 flux and soil-gas concentrations," page 191 in *The Natural Radiation Environment*, Adams, J. A. S. and Lowder, W. W., Eds. (University of Chicago Press, Chicago, Illinois).

KRUGER, J. and ANDREWS, M. (1976). "Measurement of the attachment coefficient of radon-220 decay products to monodisperse aerosols," J. Aerosol Sci. **7,** 21.

KUNZ, E. and SEVC, J. (1978). "Lung cancer mortality in uranium miners (methodological aspects)," Health Phys. **35,** 579.

KUNZ, E., SEVC, J., PLACEK, V. and HORACEK, J. (1979). "Lung cancer in man in relation to different time distribution of radiation exposure," Health Phys. **36,** 699.

LANDAHL, H. D. (1950). "On the removal of airborne droplets by the human respiratory tract: I. The lung," Bull. Math. Biophys. **12,** 43.

LANDMAN, K. A. (1982). "Diffusion of radon through cracks in a concrete slab," Health Phys. **43,** 65.

LASSEN, L. and RAU, G. (1960). "Die anlagerung radioaktiver atome an aerosole (Schwebstoffe)," Z. Physik **160,** 504.

LETOURNEAU, E., MAO, Y., MCGREGOR, R. G., SEMENCIW, R., SMITH, M. H. and WIGLE, D. T. (1983). "Lung cancer mortality and indoor radon concentrations in 18 Canadian cities," 16th Midyear Topical Meeting, Epidemiology Applied to Health Physics, Albuquerque, New Mexico, Jan. 1983.

LONDON, E. S. (1904). "Uber die physiologischen Wirkungen der Emanation des Radiums," Zntralbl. Physiol. **18,** 185.

LUNDIN, F. E., JR., WAGONER, J. K. and ARCHER, V. E. (1971). *Radon Daughter Exposure and Respiratory Cancer Quantitative and Temporal Aspects*, National Institute for Occupational Safety and Health, National Institute of Environmental Health Sciences, Joint Monograph No. 1 (National Technical Information Service, Springfield, Virginia).

LUZ, A., MULLER, W. A., GOSSNER, W. and HUG, O. (1976). "Estimation of tumor risk at low dose from experimental results after incorporation of short-lived bone-seeking alpha emitters in mice," page 171 in *Biological and Environmental Effects of Low Level Radiation. Vol. 2*, IAEA Publication STI/PUB/409 (International Atomic Energy Atomic, Vienna).

MACBETH, P. J., JENSEN, C. M., ROGERS, V. C. and OVERMEYER, R. F. (1978). *Laboratory Research on Tailings Stabilization Methods and Their Effectiveness in Radiation Containment*, U.S. Department of Energy Report GJT-21 (National Technical Information Service, Springfield, Virginia).

MARPLE, M. L. and CLEMENTS, W. E. (1978). *Contribution of Radon-222 to the Atmosphere from Inactive Uranium Mill Tailings in U.S.*, DOE Report LASL-7254 PR. (U.S. Department of Energy, Washington, D.C.).

MARTIN, D. and JACOBI, W. (1972). "Diffusion deposition of small sized particles in the bronchial tree," Health Phys. **23,** 23.

MAYS, C. W., SPIES, H. and GERSPACH, A. (1978). "Skeletal effects following Ra-224 injections into humans," Health Phys. **35,** 83.

McCULLOUGH, R., STOCKER, H., AND MAKEPEACE, C. E. (1979). "Pilot study on radon daughter exposures in Canada," page 183 in *Conference/Workshop on Lung Cancer Epidemiology and Industrial Applications of Sputum Cytology* (Colorado School of Mines Press, Golden, Colorado).

McGREGOR, R. G., VASUDEV, P., LETOURNEAU, E. G., McCULLOUGH, R. S., PRANTL, F. A. and TANAGUCHI, H. (1980). "Background concentration of radon and radon daughters in Canadian homes," Health Phys. **39,** 285.

McMICHAEL, A. J. (1976). "SMR's and the healthy worker effect: scratching beneath the surface," J. Occup. Med. **18,** 165.

MERCER, T. T. (1976). "The effect of particle size on the escape of recoiling RaB atoms from particulate surfaces," Health Phys. **31,** 173.

MERCER, T. and STOWE, W. A. (1970). "Radioactive aerosols produced by radon in room air," page 839 in *Inhaled Particles III*, Vol. II, Walton, W. H., Ed. (Unwin Bros. Ltd., Gresham Press, Old Woking, Surrey, England).

MITCHELL, J. S. (1945). *Memorandum on Some Aspects of the Biological Action of Radiations*, Special Reference to Tolerance Problems, Montreal Lab. Report HI-17 (Atomic Energy of Canada, Ltd., Montreal).

MOHNEN, V. (1967). *Investigation of the Attachment of Neutral and Electrically Charged Emanation Decay Products to Aerosols*, AERE Trans. **1106** (Doctoral Thesis).

MORGAN, K. Z. (1964). *Maximum Permissible Concentration of Radon in the Air*, Unpublished memoranda: April 1951, October 1954, cited in "The hazards of inhaling radon-222 and its short-lived daughters: A consideration of proposed maximum concentrations in air," Vol. 1, page 333 in *Radiological Health and Safety in Mining and Milling of Nuclear Materials*, IAEA Publication STI/PUB/78 (International Atomic Energy Agency, Vienna).

MORKEN, D. A. (1955a). *A Survey of the Literature on the Biological Effects of Radon and a Determination of Its Acute Toxicity*, University of Rochester Atomic Energy Project, Report UR-379 (University of Rochester, Rochester, New York).

MORKEN, D. A. (1955b). "Acute toxicity of radon," AMA Arch. Ind. Health **12,** 435.

MORKEN, D. A. (1973b). "The biological effects of radon on the lung," page 501 in *Noble Gases*, Stanley, R. E. and Moghissi, A. A., Eds., National Environmental Research Center Report CONF-730915 (National Environmental Research Center, Las Vegas, Nevada).

MORKEN, D. A. and SCOTT, J. K. (1966). *Effects on Mice of Continual Exposure to Radon and Its Decay Products on Dust*, University of Rochester Atomic Energy Project, Report UR-669 (National Technical Information Service, Springfield, Virginia).

MULLER, W. A., GOSSNER, W., HUG, O. and LUZ, A. (1978). "Late effects after incorporation of the short-lived emitters Ra-224 and Th-227 in mice," Health Phys. **35,** 33.

MYERS, D. K. and STEWART, C. G. (1979). "Some Health Aspects of Canadian Uranium Mining," page 368 in *Conference/Workshop on Lung Cancer Epidemiology and Industrial Applications of Sputum Cytology* (Colorado School of Miners Press, Golden Colorado). Also in Chalk River Laboratory Report AECL 5970.

NAS (1972). National Academy of Sciences, *The Effects on Populations of Exposure to Low Levels of Ionizing Radiation* (*BEIR II Report*). (Government Printing Office, Washington, D.C.).

NAS (1980). National Academy of Sciences, *The Effect on Populations of Exposure to Low Level of Ionizing Radiation* (*BEIR III Report*). National Academy Press. (National Academy of Science, 2101 Constitution Avenue, N.W., Washington, D.C. 20318).

NBS (1941). National Bureau of Standards, *Safe Handling of Radioactive Luminous Compounds*, National Bureau of Standards Handbook H-27 (out of print) (also NCRP Report No. 5).

NBS (1952). National Bureau of Standards, *Maximum Permissible Amount of Radioisotopes in the Human Body and Maximum Permissible Concentrations in Air and in Water*, National Bureau of Standards Handbook 52 (superseded by Handbook 69) (also NCRP Report No. 11).

NBS (1959). National Bureau of Standards, *Maximum Permissible Body Burdens and Maximum Permissible Concentrations of Radionuclides in Air and in Water for Occupational Exposure*, NBS Handbook 69; *Recommendations of the National Committee on Radiation Protection*, NCRP Report No. 22 (National Council on Radiation Protection and Measurements, Bethesda, Maryland).

NCRP (1971). National Council on Radiation Protection and Measurements, *Basic Radiation Protection Criteria*, NCRP Report No. 22 (National Council on Radiation Protection and Measurements, Bethesda, Maryland).

NCRP (1975). National Council on Radiation Protection and Measurements, *Natural Background Radiation in the United States*, NCRP Report No. 45 (National Council on Radiation Protection and Measurements, Bethesda, Maryland).

NCRP (1984). National Council on Radiation Protection and Measurements, *Exposures from the Uranium Series with Emphasis on Radon and its Daughters*, NCRP Report No. 77 (National Council on Radiation Protection and Measurements, Bethesda, Maryland).

NEA (1976). Nuclear Energy Agency, *Personal Dosimetry and Area Monitoring Suitable for Radon and Daughter Products*, Proceedings Nuclear Energy Agency Specialist Meeting (Nuclear Energy Agency, OECD, Paris).

NELSON, I. C., PARKER, H. M., CROSS, F. T., CRAIG, D. K. and STUART, B. O. (1974). *A Further Appraisal of Dosimetry Related to Uranium Mining Health Hazards*, U. S. Public Health Service Report CPE 69-1131 (National Institute of Occupational Safety and Health, Cincinnati, Ohio).

NEUFELD, J. and SNYDER, W. S. (1961). "Estimates of energy dissipation by heavy charged particles in tissue," page 35 in *Selected Topics in Radiation Dosimetry*, IAEA Publication STI/PUb/25 (International Atomic Energy Agency, Vienna).

NIEWOEHNER, D. E., KLEINERMAN, J. and RICE, B. (1974). "Pathologic changes in the peripheral airways of young cigarette smokers," N. Engl. J.

Med. **291,** 755.

NIOSH/NIEHS (1971). National Institute for Occuaptional Safety and Health/National Institute of Environmental Health Sciences, Joint Monograph No. 1, *Radon Daughter Exposure and Respiratory Cancer—Quantitative and Temporal Aspects* (National Technical Information Service, Springfield, Virginia).

NOWELL, J. A. and TYLER, W. S. (1971). "Scanning electron microscopy of the surface morphology of mammalian lungs," Am. Rev. Respir. Dis. **103,** 313.

OOSTHUIZEN, S. F., PYNE-MERCIER, W. C., FICHARDT, T. and SAVAGE, D. (1958). "Experience in radiological protection in South Africa," page 25 in *Proceedings 2nd United Nations International Conference on Peaceful Uses of Atomic Energy. Vol. 21* (United Nations, New York).

PALMER, H. E., PERKINS, R. W. and STUART, B. O. (1964). "The distribution and deposition of radon daughters attached to dust particles in the respiratory system of humans exposed to uranium mine atmospheres," Health Phys. **10,** 1129.

PARKER, H. M. (1969). "The dilemma of lung dosimetry," Health Phys. **16,** 558.

PEARSON, J. E. (1967). *Natural Environmental Radioactivity from Radon-222,* U. S. Public Health Service Report 999-RH-26 (Public Health Service, Washington, D.C.).

PHALEN, R. F., YEH, H. C., SCHUM, G. M. and RAABE, O. G. (1978). "Application of an idealized model to morphometry of the mammalian tracheobronchial tree," Anat. Rec. **190,** 167.

PORSTENDORFER, J. and MERCER, T. T. (1978). "Condensation distributions of free and attached Rn and Tn decay products in laminar aerosol flow in a cylindrical tube," J. Aerosol Sci. **9,** 283.

PORSTENDORFER, J., WICKE, A. and SCHRAUB, A. (1978). "The influence of exhalation, ventilation and deposition processes upon the concentration of radon (Rn-222), thoron (Rn-220) and their decay products in room air," Health Phys. **34,** 465.

PRADEL, J. (1973). *"Radon protection in uranium mines,"* page 605 in *Noble Gases*, Stanley, R.E. and Moghissi, A.A., Eds., National Environmental Research Center Report CONF-730915 (National Environmental Research Center, Las Vegas, Nevada).

RAABE, O. G. (1969). "Concerning in the interactions that occur between radon decay products and aerosols," Health Phys. **17,** 177.

RAABE, O. G. (1977). "Interaction between radon daughters and aerosols," page 26 in *Workshop on Dosimetry for Radon and Radon Daughters,* Turner, J. E. and Holoway, C. F., Eds. (Technical Information Service, Springfield, Virginia).

RAABE, O. G., YEH, H. C., SCHUM, G. M. and PHALEN, R. F. (1976). *Tracheobronchial Geometry: Human, Dog, Rat, Hamster,* Inhalation Toxicology Research Institute, Lovelace Foundation for Medical Education and Research Report LF-53 (National Technical Information Service, Springfield, Virginia).

RADFORD, E. P. (1981). "Radon Daughters in the Induction of Lung Cancer in Underground Miners," Banbury Report No. 9. (Cold Spring Harbor Laboratory, Cold Spring Harbor, New York).

RADFORD, E. P. and RENARD, K. G. St. Clair (1981). "Mortality Experience among Swedish Iron Miners," Presented at The International Conference on Radiation Hazards in Mining: Control, Measurement and Medical Aspects, Golden, Colorado, Oct. 5–9, 1981.

RAJEWSKY, B., SCHRAUB, A. and SCHRAUB, E. (1942a). "Uber die toxische dosis by einattmung von Ra-emanation," Naturwissenschaften **30,** 489.

RAJEWSKY, B., SCHRAUB, A. and SCHRAUB, E. (1942b). "Zur frage der toleranze-dosis bei der einatmung von Ra-em," Naturwissenschaften **30,** 733.

RAJEWSKY, B., SCHRAUB, A. and KAHLAU, G. (1943). "Experimentelle geschwulsterzeugung duch einatmung von radiumemanatory," Naturwissenschaften **31,** 170.

READ, J. and MOTTRAM, J. C. (1939). "The 'tolerance concentration' of radon in the atmosphere," Br. J. Radiol. **12,** 54.

REISSLAND, J. A., KAY, P. and DOLPHIN, G. W. (1976). "The observation and analysis of cancer deaths among classified radiation workers," Phys. Med. Biol.. **21,** 903.

RENARD, K. G. (1974). "Respiratory cancer mortality in an iron mine in northern Sweden," Ambia **3,** 67.

RICHMOND, C. R. and BOECKER, B. B. (1971). "Experimental studies," *Final Report of Subgroups IB. Interagency Uranium Mining Radiation Review Group* (U.S. Environmental Protection Agency, Rockville, Maryland).

RIEZLER, W. and SCHEPERS, H. (1961). "Ionization and energy loss of alpha particles in various gases," Ann. Phys. (Leipzig), **8,** 270.

ROESSLER, C. E., KAUTZ, R., BOLCH, W. E., JR. and WETHINGTON, J. A., JR. (1978). "The effect of mining and land reclamation on the radiological characteristics of the terrestrial environment of Florida's phosphate regions," (Presented at *Natural Radiation Environment III*, Houston, Texas, April 1978).

ROUSSEL, J., PERNOT, C., SCHOUMACHER, P., PERNOT, M. and KESSLER, Y. (1964). "Considerations statistiques sur le cancer bronchique du mineur de fer du bassin de Lorraine," J. Radiol. Electr. **45,** 541.

Royal Commission (1969). *Report Respecting Radiation. Compensation and Safety at the Fluorspar Mines* (Office of Attorney General of Newfoundland, St. Lawrence, Newfoundland).

SACCOMANNO, G., ARCHER, V. E., SAUNDERS, R. P., JAMES, L. A. and BECKLER, P. A. (1964). "Lung Cancer of Uranium Miners on the Colorado Plateau," Health Phys. **10,** 1195.

SACCOMANNO, G., ARCHER, V. E., AUERBACH, O., KUSCHNER, M., SAUNDERS, R. P. and KLEIN, M. G. (1971). "Histologic types of lung cancer among uranium miners," Cancer **27,** 515.

SACCOMANNO, G., ARCHER, V. E., AUERBACH, O., SAUNDERS, R. P. and BRENNAN, L. M. (1974). "Development of carcinoma of the lung as reflected in exfoliated cells," Cancer **33,** 256.

SACCOMANNO, G. (1981). Private communication.

SACCOMANNO, G., ARCHER, V. E., AUERBACH, O., KUSCHNER, M., EGGER, M., WOOD, S. and MICK, R. (1982). "Age Factor in Histological Type of Lung Cancer among Uranium Miners. A Preliminary Report," *Proc. Of International Conference on Radiation Hazards in Mining: Control Measurement and Medical Aspects. Golden, Colorado, Oct. 5–9, 1981 (Society of Mining Engineers, New York, New York).*

SCHLESINGER, R. B. and LIPPMAN, M. (1978). "Selective particle deposition and bronchogenic carcinoma," Environ. Research **15,** 424.

SCHLESINGER, R. B., BOHNING, D. E., CHAN, T. L. and LIPPMANN, M . (1977). "Particle deposition in a hollow cast of the human tracheobronchial tree," J. Aerosol Sci. **8,** 429.

SCHUMAN, G. (1972). "Radon isotopes and daughters in the atmosphere," Arch. Meteorol. Geophys. Bioklimatol. Ser. A: **21,** 149.

SCIOCCHETTI, G. (1976). "Assessment of airborne radioactivity in Italian mines," page 259 in *Proceedings of Nuclear Energy Agency Specialist Meeting. Personal Dosimetry and Area Monitoring Suitable for Radon and Daughter Products* (Nuclear Energy Agency, OECD, Paris).

SCOTT, J. K. (1955). *The Histopathology of Mice Exposed to Radon,* University of Rochester Atomic Energy Project Report Ur-411 (University of Rochester, Rochester, New York).

SEVC, J., KUNZ, E. and PLACEK, V. (1976). "Lung cancer in uranium miners and long term exposure to radon daughter products," Health Phys. **30,** 433.

SHAPIRO, J. (1954). *An Evaluation of the Pulmonary Radiation Dosage from Radon and Its Daughter Products,* University of Rochester Atomic Energy Projects Report Ur-298 (University of Rochester, Rochester, New York).

SHAPIRO, J. (1956). "Radiation dosage from breathing radon and its daughter products," AMA Arch. Ind. Health **14,** 169.

SHAPIRO, J. and BALE, W. F. (1953). "A partial evaluation of the hazard of radon and degradation products," page 6 in *Quarterly Technical Report for October 1, 1952,* University of Rochester Atomic Energy Project Report UR-242 (University of Rochester, Rochester, New York).

SHEARER, S. D., JR. and SILL, C. W. (1969). "Evaluation of atmospheric radon in the vicinity of uranium tailings," Health Phys. **17,** 77.

SINCLAIR, D. and HOOPES, G. S. (1975). "A novel form of diffusion battery," Am. Ind. Hyg. Assoc. J. **39042,** January 1975.

SLADE, M. B. (1979). "Radiation problems in underground mines," page 452 in *Conference/Workshop on Lung Cancer Epidemiology and Industrial Applications of Sputum Cytology* (Colorado School of Mines Press, Golden, Colorado).

SNIHS, J. O. (1973). "The significance of radon and its progeny as natural radiation sources in Sweden," page 115 in *Noble Gases,* Stanley, R. E. and Moghissi, A. A., Eds., National Environmental Research Center Report CONF-730915 (National Environmental Research Center, Las Vegas, Nevada).

SNIHS, J. O. (1975). "The approach to radon problems in non-uranium mines in Sweden," page 900 in *Proceedings of the Third International Congress of the International Radiation Protection Association*, U. S. Atomic Energy Commission Report CONF-730907 (National Technical Information Service, Springfield, Virginia).

SPIERS, F. W. (1956). "Radioactivity in man and his environment," Br. J. Radiol. **29,** 409.

SPIESS, E. S. and MAYS, C. W. (1973). "Protection effect on bone sarcoma induction by Ra-224 in children and adults," page 437 in *Radionuclide Carcinogenesis* Atomic Energy Symposium Series 29 (National Technical Information Service, Springfield, Virginia).

Statislika Centralbyron (1965). *En postenkatundersokming Varen 1963* (Utredmingsinstitutet, Stockholm) (in Swedish).

STEINHAUSLER, F., HOFMAN, W., POHL, E. and POHL-RULING, J. (1980). "Local and temporal distribution pattern of radon and daughters in an urban environment and determination of organ dose frequency distributions with demoscopical method," page 1145 in *The Natural Radiation Environment III* (National Technical Information Service, Springfield, Virginia).

STEWART, C. C. and SIMPSON, S. D. (1964). "The hazards of inhaling radon-222 and its short-lived daughters: A consideration of proposed maximum concentrations in air," page 333 in *Radiological Health and Safety in Mining and Milling Nuclear Materials*, IAEA Publication STI/PUB/78 (International Atomic Energy Agency, Vienna).

STRANDEN, E. (1980). "Radon in dwellings and lung cancer. A discussion," Health Phys. **38,** 301.

STRANDEN, E., BERTEIG, L. and UGLETVEIT, F. (1979). "A study on radon in dwellings," Health Phys. **36,** 413.

STUART, B. O. (1978). *Inhaled Radon Daughters and Uranium Ore Dust in Rodents*, Pacific Northwest Laboatory Annual Report for 1977 to the DOE Assistant Secretary for the Environment, Report PNL-2500 pt. 1 (National Technical Information Service, Springfield, Virginia).

STURGESS, J. M. (1977). "The mucous lining of major bronchi in the rabbit lung," Am. Rev. Respir. Dis. **115,** 819.

SWEJEMARK, G. A. (1980). "Radon in Dwelling in Sweden," in *The Natural Radiation Environment III*, Gesell, T.F. and Lowder, W.M., Eds. (Technical Information Center, U. S. Department of Energy).

TANNER, A. B. (1964). "Radon migration in the ground A review," in *The Natural Radiation Environment*, Adams, J.A.S. and Lowder, W.M., Eds. (University of Chicago Press, Chicago).

THOMAS, J. W. and HINCHLIFFE, L. E. (1972). "Filtration of 0.001 μm particles by wire screens," J. Aerosol Sci. **3,** 387.

TRI-C (1953). *Tripartitie Conference on Permissible Doses* (Arden House, Harriman, New York).

TUREKIAN, K. K., NOZAKI, Y. and BENNINGER, L. K. (1977). "Geochemistry of atmospheric radon and decay products," Ann. Rev. Earth Planet Sci. **5,** 227.

TURNER, J. E., HOLOWAY, C. F. and LOEBL, A. S. (1977). *Workshop on Dosimetry for Radon and Radon Daughters*, Oak Ridge National Laboratory

Report ORNL-5348 (National Technical Information Service, Springfield, Virginia).

UNSCEAR (1972). United Nations Scientific Committee on the Effects of Atomic Radiation. *Ionizing Radiation: Levels and Effects* (United Nations, New York).

UNSCEAR (1977). United Nations Scientific Committee on the Effects of Atomic Radiation. *Sources and Effects of Ionizing Radiation* (United Nations, New York).

USPHS (1957). United States Public Health Service, *Control of Radon and Daughters in Uranium Mines and Calculations on Biologic Effects*, USPHS Publication No. 494 (Government Printing Office, Washington, D.C.).

USPHS (1961). United States Public Health Service, *Governor's Conference on Health Hazards in Uranium Mines, A summary Report*, USPHS Publication No. 843 (Public Health Service, Bureau of State Services, Washington, D.C.).

VAN AS, A. (1977). "Pulmonary airway clearance mechanisms: A reappraisal," Am. Rev. Respir. Dis. **115,** 721.

VAN AS, A. and WEBSTER, I. (1974). "The morphology of mucus in mammalian pulmonary airways," Environ. Res. **7,** 1.

WAGONER, J. K., MILLER, R. W., LUNDIN, F. E., JR, FRAUMENI, J. F. JR. and HAIJ, M. E. (1963). "Unusual cancer mortality among a group of undergroup metal miners," N. Engl. J. Med. **269,** 284.

WALSH, P. J. (1970). "Radiation dose to the respiratory tract of uranium miners," Environ. Res. **3,** 14.

WALSH, P. J. (1971). "Relationship of experimental to empirical findings and theoretical dose calculations," in *Final Report of Subgroup IB. Interagency Uranium Mining Radiation Review Group* (Environmental Protection Agency, Rockville, Maryland).

WALSH, P. J. (1979). "Dose conversion factors for radon daughters," Health Phys. **36,** 601.

WANNER, A. (1977). "Clinical aspects of mucociliary transport," Am. Rev. Respir. Dis. **116,** 73.

WAXWEILLER, R. J., WAGONER, J. K. and ARCHER, V. E. (1973). "Mortality of potash workers," J. Occup. Med. **15,** 486.

WAXWEILLER, R. J. (1981). "Updated Mortality Analysis of U.S. Uranium Miner Study Group," Proc. of International Conference on Radiation Hazards in Mining: Control. Measurement and Medical Aspects. Golden, Colorado, Oct. 5–9, 1981 (Society of Mining Engineers, New York, New York).

WEBSTER, I. and BASSON, J. K. (1969). "Bronchogenic cancer in South African gold miners," presented at the *International Conference on Pneumocaniosis,* Johannesburg.

WEIBEL, E. R. (1963). *Morphometry of the Human Lung* (Academic Press, New York).

WHITTEMORE, A. S. and MCMILLAN, A. (1983). "Lung cancer mortality among U. S. uranium miners; A reapprisal," J. Nat. Can. Inst. **71,** 489.

WILKENING, M. H. and CLEMENTS, W. E. (1975). "Radon-222 from the ocean surface," J. Geophys. Res. **80,** 3828.

WILKENING, M. H., CLEMENTS, W. E. and STANLEY, D. (1972). "Radon-222 flux measurements in widely separated regions," page 717 in *Natural Radiation in the Environment II,* Adams, J.A.S., Lowder, W.M. and Gesel, T.F., Eds., U.S. Energy Research and Development Administration Report CONF-720805-P2 (National Technical Information Service, Springfield, Virginia).

WHO (1978). World Health Organization, *World Health Statistics Annual.* (Geneva, Switzerland).

WOODARD, H. Q. (1980). *Radiation Carcinogenesis in Man: A Critical Review,* U.S. DOE Report EML-380 (National Technical Information Service, Springfield Virginia).

WRIGHT, E. S. and COUVES, C. M. (1977). "Radiation-induced carcinoma of the lung the St. Lawrence tragedy," J. Thorac. Cardiovasc. Surg. **74,** 495.

YONEDA, K. (1976). "Mucous blanket of rat bronchus," Am. Rev. Respir. Dis. **114,** 837.

YEH, H. C. and SCHUM, M. (1980). "Models of human lung airways and their application to inhaled particle deposition," Bull. Math. Biol. **42,** 461.

The NCRP

The National Council on Radiation Protection and Measurements is a nonprofit corporation chartered by Congress in 1964 to:

1. Collect, analyze, develop, and disseminate in the public interest information and recommendations about (a) protection against radiation and (b) radiation measurements, quantities, and units, particularly those concerned with radiation protection;
2. Provide a means by which organizations concerned with the scientific and related aspects of radiation protection and of radiation quantities, units, and measurements may cooperate for effective utilization of their combined resources, and to stimulate the work of such organizations;
3. Develop basic concepts about radiation quantities, units, and measurements, about the application of those concepts, and about radiation protection;
4. Cooperate with the International Commission on Radiological Protection, the International Commission on Radiation Units and Measurements, and other national and international organizations, governmental and private, concerned with radiation quantities, units, and measurements and with radiation protection.

The Council is the successor to the unincorporated association of scientists known as the National Committee on Radiation Protection and Measurements and was formed to carry on the work begun by the Committee.

The Council is made up of the members and the participants who serve on the eighty-one Scientific Committees of the Council. The Scientific Committees, composed of experts having detailed knowledge and competence in the particular area of the Committee's interest, draft proposed recommendations. These are then submitted to the full membership of the Council for careful review and approval before being published.

The following comprise the current officers and membership of the Council:

Officers

President	WARREN K. SINCLAIR
Vice President	S. JAMES ADELSTEIN
Secretary and Treasurer	W. ROGER NEY
Assistant Secretary	EUGENE R. FIDELL
Assistant Treasurer	JAMES F. BERG

Members

SEYMOUR ABRAHAMSON
S. JAMES ADELSTEIN
ROY E. ALBERT
PETER R. ALMOND
EDWARD L. ALPEN
JOHN A. AUXIER
WILLIAM J. BAIR
JOHN D. BOICE, JR.
VICTOR P. BOND
ROBERT L. BRENT
ANTONE BROOKS
REYNOLD F. BROWN
THOMAS F. BUDINGER
MELVIN W. CARTER
GEORGE W. CASARETT
RANDALL S. CASWELL
ARTHUR B. CHILTON
GERALD D. DODD
PATRICIA W. DURBIN
JOE A. ELDER
MORTIMER M. ELKIND
THOMAS S. ELY
EDWARD R. EPP
R. J. MICHAEL FRY
ROBERT A. GOEPP
ROBERT O. GORSON
ARTHUR W. GUY
ERIC J. HALL
NAOMI H. HARLEY
JOHN W. HEALY
JOHN M. HESLEP
SEYMOUR JABLON
DONALD G. JACOBS
A. EVERETTE JAMES, JR.
BERND KAHN
JAMES G. KEREIAKES
CHARLES E. LAND
GEORGE R. LEOPOLD
THOMAS A. LINCOLN
RAY D. LLOYD
ARTHUR C. LUCAS
CHARLES W. MAYS
ROGER O. MCCLELLAN
JAMES E. MCLAUGHLIN
BARBARA J. MCNEIL
THOMAS F. MEANEY
CHARLES B. MEINHOLD
MORTIMER L. MENDELSOHN
WILLIAM A. MILLS
DADE W. MOELLER
A. ALAN MOGHISSI
ROBERT D. MOSELEY, JR.
JAMES V. NEEL
WESLEY NYBORG
MARY E. O'CONNOR
FRANK L. PARKER
ANDREW K. POZNANSKI
NORMAN C. RASMUSSEN
WILLIAM C. REINIG
CHESTER R. RICHMOND
JAMES S. ROBERTSON
LEONARD A. SAGAN
WILLIAM J. SCHULL
GLENN E. SHELINE
ROY E. SHORE
WARREN K. SINCLAIR
LEWIS V. SPENCER
JOHN B. STORER
ROY C. THOMPSON
JOHN E. TILL
ARTHUR C. UPTON
GEORGE L. VOELZ
EDWARD W. WEBSTER
GEORGE M. WILKENING
H. RODNEY WITHERS

Honorary Members

LAURISTON S. TAYLOR, *Honorary President*

EDGAR C. BARNES
AUSTIN M. BRUES
FREDERICK P. COWAN
JAMES F. CROW
MERRIL EISENBUD
ROBLEY D. EVANS
RICHARD F. FOSTER
HYMER L. FRIEDELL
JOHN H. HARLEY
LOUIS N. HEMPELMANN, JR.
PAUL C. HODGES
GEORGE V. LEROY
WILFRID B. MANN
KARL Z. MORGAN
RUSELL H. MORGAN
ROBERT J. NELSEN
HERBERT M. PARKER
HARALD H. ROSSI
WILLIAM G. RUSSELL
JOHN H. RUST
EUGENE L. SAENGER
J. NEWELL STANNARD
HAROLD O. WYCKOFF

Currently, the following subgroups are actively engaged in formulating recommendations:

SC-1: Basic Radiation Protection Criteria
SC-3: Medical X-Ray, Electron Beam and Gamma-Ray Protection for Energies Up to 50 MeV (Equipment Performance and Use)
SC-16: X-Ray Protection in Dental Offices
SC-18: Standards and Measurements of Radioactivity for Radiological Use
SC-38: Waste Disposal
    Task Group on Krypton-85
    Task Group on Carbon-14
    Task Group on Disposal of Accident Generated Waste Water
    Task Group on Disposal of Low-Level Waste
    Task Group on the Actinides
    Task Group on Xenon
SC-40: Biological Aspects of Radiation Protection Criteria
    Task Group on Atomic Bomb Survivor Dosimetry
    Subgroup on Biological Aspects of Dosimetry of Atomic Bomb Survivors
SC-42: Industrial Applications of X Rays and Sealed Sources
SC-44: Radiation Associated with Medical Examinations
SC-45: Radiation Received by Radiation Employees
SC-46: Operational Radiation Safety
    Task Group 1 on Warning and Personnel Security Systems
    Task Group 2 on Uranium Mining and Milling—Radiation Safety Program
    Task Group 3 on ALARA for Occupationally Exposed Individuals in Clinical Radiology
    Task Group 4 on Calibration of Instrumentation
    Task Group 5
SC-47: Instrumentation for the Determination of Dose Equivalent
SC-48: Apportionment of Radiation Exposure
SC-52: Conceptual Basis of Calculations of Dose Distributions
SC-53: Biological Effects and Exposure Criteria for Radiofrequency Electomagnetic Radiation
SC-54: Bioassay for Assessment of Control of Intake of Radionuclides

SC-55: Experimental Verification of Internal Dosimetry Calculations
SC-57: Internal Emitter Standards
Task Group 2 on Respiratory Tract Model
Task Group 3 on General Metabolic Models
Task Group 4 on Radon and Daughters
Task Group 6 on Bone Problems
Task Group 7 on Thyroid Cancer Risk
Task Group 8 on Leukemia Risk
Task Group 9 on Lung Cancer Risk
Task Group 10 on Liver Cancer Risk
Task Group 11 on Genetic Risk
Task Group 12 on Strontium
Task Group 13 on Neptunium
SC-59: Human Radiation Exposure Experience
SC-60: Dosimetry of Neutrons from Medical Accelerators
SC-61: Radon Measurements
SC-62: Priorities for Dose Reduction Efforts
SC-63: Control of Exposure to Ionizing Radiation from Accident or Attack
SC-64: Radionuclides in the Environment
Task Group 5 on Public Exposure to Nuclear Power
Task Group 6 on Screening Models
Task Group 7 on Soil Contamination
SC-65: Quality Assurance and Accuracy in Radiation Protection Measurements
SC-67: Biological Effects of Magnetic Fields
SC-68: Microprocessors in Dosimetry
SC-69: Efficacy Studies
SC-70: Quality Assurance and Measurement in Diagnostic Radiology
SC-71: Radiation Exposure and Potentially Related Injury
SC-72: Radiation Protection in Mammography
SC-74: Radiation Received in the Decontamination of Nuclear Facilities
SC-75: Guidance on Radiation Received in Space Activities
SC-76: Effects of Radiation on the Embryo-Fetus
SC-77: Guidance on Occupational and Public Exposure Resulting from Diagnostic Nuclear Medicine Procedures
SC-78: Practical Guidance on the Evaluation of Human Exposures to Radiofrequency Radiation
SC-79: Extremely Low-Frequency Electric and Magnetic Fields
SC-80: Radiation Biology of the Skin (Beta-Ray Dosimetry)
SC-81: Assessment of Exposure from Therapy
Committee on Public Education
Ad Hoc Committee on Policy in Regard to the International System of Units
Ad Hoc Committee on Comparison of Radiation Exposures
Study Group on Acceptable Risk (Nuclear Waste)
Study Group on Comparative Risk
Task Group on Comparative Carcinogenicity of Pollutant Chemicals
Task Force on Occupational Exposure Levels

In recognition of its responsibility to facilitate and stimulate cooperation among organizations concerned with the scientific and related aspects of radiation protection and measurement, the Council has created a category of NCRP Collaborating Organizations. Organizations or groups of organizations that are national or international in scope and are concerned with scientific problems involving radiation quantities, units, measurements and effects, or radiation protection may be admitted to collaborating status by the Council. The present Collaborating Organizations with which the NCRP maintains liaison are as follows:

American Academy of Dermatology
American Association of Physicists in Medicine
American College of Nuclear-Physicians
American College of Radiology
American Dental Association
American Industrial Hygiene Association
American Institute of Ultrasound in Medicine
American Insurance Association
American Medical Association
American Nuclear Society
American Occupational Medical Association
American Podiatry Association
American Public Health Association
American Radium Safety
American Roentgen Ray Society
American Society of Radiologic Technologists
American Society for Therapeutic Radiology and Oncology
Association of University Radiologists
Atomic Industrial Forum
Bioelectromagnetics Society
College of American Pathologists
Federal Communications Commission
Federal Emergency Management Agency
Genetics Society of America
Health Physics Society
National Bureau of Standards
National Electrical Manufacturers Association
Radiation Research Society
Radiological Society of North America
Society of Nuclear Medicine
United States Army
United States Air Force
United States Department of Energy
United States Department of Labor
United States Environmental Protection Agency
United States Navy
United States Nuclear Regulatory Commission
United States Public Health Service

The NCRP has found its relationships with these organizations to be extremely valuable to continue progress in its program.

Another aspect of the cooperative efforts of the NCRP relates to the special liaison relationships established with various governmental organizations that have an interest in radiation protection and measurements. This liaison relationship provides: (1) an opportunity for participating organizations to designate an individual to provide liaison between the organization and the NCRP; (2) that the individual designated will receive copies of draft NCRP reports (at the time that these are submitted to the members of the Council) with an invitation to comment, but not vote; and (3) that new NCRP efforts might be discussed with liaison individuals as appropriate, so that they might have an opportunity to make suggestions on new studies and related matters. The following organizations participate in the special liaison program:

Defense Nuclear Agency
Federal Emergency Management Agency
National Bureau of Standards
Office of Science and Technology Policy
Office of Technology Assessment
United States Air Force
United States Army
United States Coast Guard
United States Department of Energy
United States Department of Health and Human Services
United States Department of Labor
United States Department of Transportation
United States Environmental Protection Agency
United States Navy
United States Nuclear Regulatory Commission

The NCRP values highly the participation of these organizations in the liaison program.

The Council's activities are made possible by the voluntary contribution of time and effort by its members and participants and the generous support of the following organizations:

Alfred P. Sloan Foundation
Alliance of American Insurers
American Academy of Dental Radiology
American Academy of Dermatology
American Association of Physicists in Medicine
American College of Radiology
American College of Radiology Foundation
American Dental Association
American Industrial Hygiene Association
American Insurance Association
American Medical Association
American Nuclear Society
American Occupational Medical Association
American Osteopathic College of Radiology
American Podiatry Association
American Public Health Association
American Radium Society
American Roentgen Ray Society
American Society of Radiologic Technologists
American Society of Therapeutic Radiologists
American Veterinary Medical Association
American Veterinary Radiology Society
Association of University Radiologists
Atomic Industrial Forum
Battelle Memorial Institute
Bureau of Radiological Health
College of American Pathologists
Commonwealth of Pennsylvania
Defense Nuclear Agency
Edison Electric Institute
Edward Mallinckrodt, Jr. Foundation
Electric Power Research Institute
Federal Emergency Management Agency
Florida Institute of Phosphate Research
Genetics Society of America
Health Physics Society
James Picker Foundation
National Association of Photographic Manufacturers
National Bureau of Standards
National Cancer Institute
National Electrical Manufacturers Association
Radiation Rasearch Society
Radiological Society of North America
Society of Nuclear Medicine
United States Department of Energy
United States Department of Labor
United States Environmental Protection Agency
United States Navy
United States Nuclear Regulatory Commission

To all these organizations the Council expresses its profound appreciation for their support.

Initial funds for publication of NCRP reports were provided by a grant from the James Picker Foundation and for this the Council wishes to express it deep appreciation.

The NCRP seeks to promulgate information and recommendations based on leading scientific judgement on matters of radiation protection and measurement and to foster cooperation among organizations concerned with these matters. These efforts are intended to serve the public interest and the Council welcomes comments and suggestions on its reports or activities from those interested in its work.

NCRP Publications

NCRP publications are distributed by the NCRP Publications' office. Information on prices and how to order may be obtained by directing an inquiry to:

NCRP Publications
7910 Woodmont Ave, Suite 1016
Bethesda, Md. 20814

The currently available publications are listed below.

Proceedings of the Annual Meeting

No.	Title
1	*Perceptions of Risk,* Proceedings of the Fifteenth Annual Meeting, Held on March 14–15, 1979 (Including Taylor Lecture No. 3) (1980)
2	*Quantitative Risk in Standards Setting,* Proceedings of the Sixteenth Annual Meeting Held on April 2–3, 1980 (Including Taylor Lecture No. 4) (1981)
3	*Critical Issues in Setting Radiation Dose Limits,* Proceedings of the Seventeenth Annual Meeting, Held on April 8–9, 1981 (Including Taylor Lecture No. 5) (1982)
4	*Radiation Protection and New Medical Diagnostic Procedures,* Proceedings of the Eighteenth Annual Meeting, Held on April 6–7, 1982 (Including Taylor Lecture No. 6) (1983)
5	*Environmental Radioactivity,* Proceedings of the Nineteenth Annual Meeting, held on April 6–7, 1983 (Including Taylor Lecture No. 7) (1984)

Symposium Proceedings

The Control of Exposure of the Public to Ionizing Radiation in the Event of Accident or Attack, Proceedings of a Symposium held on April 27–29, 1981. (1982).

Lauriston S. Taylor Lectures

No.	Title and Author
1	*The Squares of the Natural Numbers in Radiation Protection* by Herbert M. Parker (1977)

2 *Why be Quantitative About Radiation Risk Estimates?* by Sir Edward Pochin (1978)

3 *Radiation Protection—Concepts and Trade Offs* by Hymer L. Friedell (1979) [Available also in *Perceptions of Risk*, see above]

4 *From "Quantity of Radiation" and "Dose" to "Exposure" and "Absorbed Dose"—An Historical Review* by Harold O. Wyckoff (1980) [Available also in *Quantitative Risks in Standards Setting*, see above]

5 *How Well Can We Assess Genetic Risk? Not Very* by James F. Crow (1981) [Available also in *Critical Issues in Setting Radiation Dose Limits*, see above]

6 *Ethics, Trade-offs and Medical Radiation* by Eugene L. Saenger (1982) [Available also in *Radiation Protection and New Medical Diagnostic Approaches*, see above.]

7 *The Human Environment-Past, Present and Future* by Merril Eisenbud (1983) [Available also in *Environmental Radioactivity*, see above.]

NCRP Reports

No.	Title
8	*Control and Removal of Radioactive Contamination in Laboratories* (1951)
9	*Recommendations for Waste Disposal of Phosphorus-32 and Iodine-131 for Medical Users* (1951)
12	*Recommendations for the Disposal of Carbon-14 Wastes* (1953)
16	*Radioactive Waste Disposal in the Ocean* (1954)
22	*Maximum Permissible Body Burdens and Maximum Permissible Concentrations of Radionuclides in Air and in Water for Occupational Exposure* (1959) [Includes Addendum 1 issued in August 1963]
23	*Measurement of Neutron Flux and Spectra for Physical and Biological Applications* (1960)
25	*Measurement of Absorbed Dose of Neutrons and Mixtures of Neutrons and Gamma Rays* (1961)
27	*Stopping Powers for Use with Cavity Chambers* (1961)
30	*Safe Handling of Radioactive Materials* (1964)
32	*Radiation Protection in Educational Institutions* (1966)
33	*Medical X-Ray and Gamma-Ray Protection for Energies Up to 10 MeV—Equipment Design and Use* (1968)
35	*Dental X-Ray Protection* (1970)

61 *Radiation Safety Training Criteria for Industrial Radiography* (1978)
62 *Tritium in the Environment* (1979)
63 *Tritium and Other Radionuclide Labeled Organic Compounds Incorporated in Genetic Material* (1979)
64 *Influence of Dose and Its Distribution in Time on Dose-Response Relationships for Low-LET Radiations* (1980)
65 *Management of Persons Accidentally Contaminated with Radionuclides* (1980)
66 *Mammography* (1980)
67 *Radiofrequency Electromagnetic Fields—Properties, Quantities and Units, Biophysical Interaction, and Measurements* (1981)
68 *Radiation Protection in Pediatric Radiology* (1981)
69 *Dosimetry of X-Ray and Gamma-Ray Beams for Radiation Therapy in the Energy Range 10 keV to 50 MeV* (1981)
70 *Nuclear Medicine-Factors Influencing the Choice and Use of Radionuclides in Diagnosis and Therapy* (1982)
71 *Operational Radiation Safety—Training* (1983)
72 *Radiation Protection and Measurement for Low Voltage Neutron Generators* (1983)
73 *Protection in Nuclear Medicine and Ultrasound Diagnostic Procedures in Children* (1983)
74 *Biological Effects of Ultrasound: Mechanisms and Clinical Implications* (1983)
75 *Iodine-129: Evaluation of Releases from Nuclear Power Generation* (1983)
76 *Radiological Assessment: Predicting the Transport, Bioaccumulation, and Uptake by Man of Radio-nuclides Released to the Environment* (1984)
77 *Exposures from the Uranium Series with Emphasis on Radon and its Daughters* (1984)
78 *Evaluation of Occupational and Environmental Exposures to Radon and Radon Daughters in the United States* (1984)

Binders for NCRP Reports are available. Two sizes make it possible to collect into small binders the "old series" of reports (NCRP Reports Nos. 8–30) and into large binders the more recent publications (NCRP Reports Nos. 32–74). Each binder will accommodate from five to seven reports. The binders carry the identification "NCRP Reports" and come with label holders which permit the user to attach labels showing the reports contained in each binder.

The following bound sets of NCRP Reports are also available:

Volume I. NCRP Reports Nos. 8, 9, 12, 16, 22
Volume II. NCRP Reports Nos. 23, 25, 27, 30
Volume III. NCRP Reports Nos. 32, 33, 35, 36, 37
Volume IV. NCRP Reports Nos. 38, 39, 40, 41
Volume V. NCRP Reports Nos. 42, 43, 44, 45, 46
Volume VI. NCRP Reports Nos. 47, 48, 49, 50, 51
Volume VII. NCRP Reports Nos. 52, 53, 54, 55, 56, 57
Volume VIII. NCRP Report No. 58
Volume IX. NCRP Reports Nos. 59, 60, 61, 62, 63
Volume X. NCRP Reports Nos. 64, 65, 66, 67
Volume XI. NCRP Reports Nos. 68, 69, 70, 71, 72

(Titles of the individual reports contained in each volume are given above).

The following NCRP Reports are now superseded and/or out of print:

No.	Title
1	*X-Ray Protection* (1931). [Superseded by NCRP Report No. 3]
2	*Radium Protection* (1934). [Superseded by NCRP Report No. 4]
3	*X-Ray Protection* (1936). [Superseded by NCRP Report No. 6]
4	*Radium Protection* (1938). [Superseded by NCRP Report No. 13]
5	*Safe Handling of Radioactive Luminous Compounds* (1941). [Out of Print]
6	*Medical X-Ray Protection Up to Two Million Volts* (1949). [Superseded by NCRP Report No. 18]
7	*Safe Handling of Radioactive Isotopes* (1949). [Superseded by NCRP Report No. 30]
10	*Radiological Monitoring Methods and Instruments* (1952). [Superseded by NCRP Report No. 57]
11	*Maximum Permissible Amounts of Radioisotopes in the Human Body and Maximum Permissible Concentrations in Air and Water* (1953). [Superseded by NCRP Report No. 22]
13	*Protection Against Radiations from Radium, Cobalt-60 and Cesium-137* (1954). [Superseded by NCRP Report No. 24]

14 *Protection Against Betatron—Synchrotron Radiations Up to 100 Million Electron Volts* (1954). [Superseded by NCRP Report No. 51]

15 *Safe Handling of Cadavers Containing Radioactive Isotopes* (1953). [Superseded by NCRP Report No. 21]

17 *Permissible Dose from External Sources of Ionizing Radiation* (1954) including *Maximum Permissible Exposure to Man, Addendum to National Bureau of Standards Handbook 59* (1958). [Superseded by NCRP Report No. 39]

18 *X-Ray Protection* (1955). [Superseded by NCRP Report No. 26

19 *Regulation of Radiation Exposure by Legislative Means* (1955). [Out of print]

20 *Protection Against Neutron Radiation Up to 30 Million Electron Volts* (1957). [Superseded by NCRP Report No. 38]

21 *Safe Handling of Bodies Containing Radioactive Isotopes* (1958). [Superseded by NCRP Report No. 37]

24 *Protection Against Radiations from Sealed Gamma Sources* (1960). [Superseded by NCRP Report Nos. 33, 34, and 40]

26 *Medical X-Ray Protection Up to Three Million Volts* (1961). [Superseded by NCRP Report Nos. 33, 34, 35, and 36]

28 *A Manual of Radioactivity Procedures* (1961). [Superseded by NCRP Report No. 58]

29 *Exposure to Radiation in an Emergency* (1962). [Superseded by NCRP Report No. 42]

31 *Shielding for High Energy Electron Accelerator Installations* (1964). [Superseded by NCRP Report No. 51]

34 *Medical X-Ray and Gamma-Ray Protection for Energies Up to 10 MeV—Structural Shielding Design and Evaluation* (1970). [Superseded by NCRP Report No. 49]

Other Documents

The following documents of the NCRP were published outside of the NCRP Reports series:

"Blood Counts, Statement of the National Committee on Radiation Protection," Radiology 63, 428 (1954)

"Statements on Maximum Permissible Dose from Television Receivers and Maximum Permissible Dose to the Skin of the Whole Body," Am. J.

Roentgenol., Radium Ther. and Nucl. Med. 84, 152 (1960) and Radiology 75, 122 (1960)

X-Ray Protection Standards for Home Television Receivers, Interim Statement of the National Council on Radiation Protection and Measurements (National Council on Radiation Protection and Measurements, Washington, 1968)

Specification of Units of Natural Uranium and Natural Thorium (National Council on Radiation Protection and Measurements, Washington, 1973)

NCRP Statement on Dose Limit for Neutrons (National Council on Radiation Protection and Measurements, Washington, 1980)

Krypton-85 in the Atmosphere—With Specific Reference to the Public Health Significance of the Proposed Controlled Release at Three Mile Island (National Council on Radiation Protection and Measurements, Washington, 1980)

Preliminary Evaluation of Criteria For the Disposal of Transuranic Contaminated Waste (National Council on Radiation Protection and Measurements, Bethesda, Md, 1982)

Copies of the statements published in journals may be consulted in libraries. A limited number of copies of the remaining documents listed above are available for distribution by NCRP Publications.

Index